Mathematical Magnetohydrodynamics

Synthesis Lectures on Mechanical Engineering

Synthesis Lectures on Mechanical Engineering series publishes 60–150 page publications pertaining to this diverse discipline of mechanical engineering. The series presents Lectures written for an audience of researchers, industry engineers, undergraduate and graduate students.

Additional Synthesis series will be developed covering key areas within mechanical engineering.

Mathematical Magnetohydrodynamics
Nikolas Xiros
2017

Microcontroller Education: Do it Yourself, Reinvent the Wheel, Code to Learn
Dimosthenis E. Bolanakis
2017

Solving Practical Engineering Mechanics Problems: Statics
Sayavur I. Bakhtiyarov
2017

Unmanned Aircraft Design: A Review of Fundamentals
Mohammad Sadraey
2017

Introduction to Refrigeration and Air Conditioning Systems: Theory and Applications
Allan Kirkpatrick
2017

Resistance Spot Welding: Fundamentals and Applications for the Automotive Industry
Menachem Kimchi and David H. Phillips
2017

MEMS Barometers Toward Vertical Position Detecton: Background Theory, System Prototyping, and Measurement Analysis
Dimosthenis E. Bolanakis
2017

Vehicle Suspension System Technology and Design
Avesta Goodarzi and Amir Khajepour
2017

Engineering Finite Element Analysis
Ramana M. Pidaparti
2017

Mathematical Magnetohydrodynamics

Nikolas Xiros

ISBN: 978-3-031-79600-5   paperback
ISBN: 978-3-031-79601-2   ebook
ISBN: 978-3-031-79602-9   hardcover

DOI 10.1007/978-3-031-79601-2

A Publication in the Springer series

SYNTHESIS LECTURES ON MECHANICAL ENGINEERING

Lecture #10
Series ISSN
Print 2573-3168   Electronic 2573-3176

Mathematical Magnetohydrodynamics

Nikolas Xiros

University of New Orleans

SYNTHESIS LECTURES ON MECHANICAL ENGINEERING #10

ABSTRACT

Fundamentals of mathematical magnetohydrodynamics (MHD) start with definitions of major variables and parameters in MHD fluids (also known as MHD media) and specifically plasmas encountered in nature as well as in engineering sytems, e.g., metallurgy or thermonuclear fusion power. Then collisions of fluids in such fluids are examined as well as motion of individual particles. Then the basic principles of MHD fluids are introduced along with transport phenomena, medium boundaries, and surface interactions. Then, waves and resonances of all sorts in MHD media are presented. The account concludes with the description of main MHD fluid types including plasma in fusion power generation.

KEYWORDS

partial differential equations, instabilities, plasma, controlled thermonuclear nuclear fusion, magnetohydrodynamics (MHD)

Contents

1   Plasma Definition and Classification 1
- 1.1   Definitions 1
- 1.2   Maxwellian Temperature Distribution 2
- 1.3   Debye Length... 3
- 1.4   Plasma Frequency... 4
- 1.5   Classification of Plasmas 5

2   Collisions in Plasmas 7
- 2.1   General Definitions 7
- 2.2   Binary Elastic Collision Kinematics 8
- 2.3   Differential Cross Section 9
- 2.4   Momentum Transfer 10
- 2.5   Coulomb Collisions 11
- 2.6   Collisions of Neutrals 13
- 2.7   Resonant Charge Transfer 14
- 2.8   Polarization Scattering 16
- 2.9   Electron Elastic Scattering at Neutrals 17
- 2.10 Electron Impact Ionization 17
- 2.11 Electron Impact Dissociation 19
- 2.12 Electron Impact Excitation 21
- 2.13 Penning Ionization 22
- 2.14 Chemical Reactions 22

3   Motion of Charged Particles 25
- 3.1   Equation of Motion 25
- 3.2   Constant Magnetic Field 25
- 3.3   Constant Electric and Magnetic Fields 25
- 3.4   Inhomogeneous Magnetic Field 27
- 3.5   Gravitation and Magnetic Field............................. 30
- 3.6   Drifts and Instabilities 30

	3.7	Time-dependent Magnetic Field	31
	3.8	Time-dependent Electric Field	34
	3.9	Adiabatic Invariants	35
4		**Plasma as a Fluid**	**37**
	4.1	Distribution Function and Moments	37
	4.2	Particle, Momentum, and Energy Balance	39
	4.3	Drifts in Fluid Description	41
5		**Transport**	**45**
	5.1	Drift and Diffusion	45
	5.2	Transport of Neutrals	46
	5.3	Ambipolar Diffusion	49
	5.4	Diffusion in a Magnetic Field	50
	5.5	Plasma Resistivity	52
	5.6	Electrical Plasma Heating	54
6		**Plasma Boundary**	**57**
	6.1	Electrostatic Sheath	57
	6.2	Presheath	59
	6.3	Potential, Flux, Ion Energy	60
	6.4	Negatively Biased Eelectrode	61
	6.5	Collisional Sheath	66
	6.6	Electrostatic Probe	68
7		**Plasma-surface Interaction**	**73**
	7.1	Ion Implantation and Reemission	73
	7.2	Collision Cascade	81
	7.3	Radiation Damage	82
	7.4	Sputtering	83
	7.5	Chemical Sputtering	85
	7.6	Surface Reactions	86
	7.7	Secondary Electron Emission	89

8 Particle Waves and Resonances **93**

    8.1   Electron Oscillations 93

    8.2   Electron Waves 94

    8.3   Ion Waves 95

    8.4   Electron Oscillations in Magnetic Fields 96

    8.5   Ion Waves in Magnetic Fields 97

9 Electromagnetic Waves **101**

    9.1   Non-magnetized Plasma 101

    9.2   Magnetized Plasma 104

10 Plasma Modeling **111**

    10.1 Global model 111

    10.2 Reactive Plasmas 115

    10.3 Fluid Modeling 117

    10.4 Particle-in-Cell Computer Simulation 118

11 Low-temperature DC Plasma **123**

    11.1 Breakdown 123

    11.2 Regimes of Operation 125

    11.3 DC Magnetron Discharge 127

12 Low-temperature RF Plasmas **131**

    12.1 Capacitively Coupled RF Discharge 131

    12.2 Ion Energy Distribution 137

13 Magnetic Confinement Nuclear Fusion Plasma **143**

    13.1 Fusion Reactions 143

    13.2 Ignition 144

    13.3 Machine Concepts 146

    13.4 Transport 150

References **153**

Author's Biography **155**

CHAPTER 1

Plasma Definition and Classification

1.1 DEFINITIONS

A **plasma**, the so-called 4th physical condition of matter, is defined as a (partly) ionized gas, which is neutral in average ("**quasineutral**") and exhibits **collective** properties (the density has to be high enough).

Thus, the plasma consists in most cases of neutral atoms or molecules, positive ions and electrons, which interact by **collisions**. The most efficient mechanism of **ionization** is the collision between electrons and neutrals. For this purpose, a fraction of the electrons has to have kinetic energies which exceed the ionization potential, i.e., above a few eV, so that the electrons have to be rather "hot."

In a stationary plasma, the rate of ionization has to be compensated by a **loss** of charged particles to the environment (wall—see Fig. 1.1), if recombination is neglected. This defines a mean residence time (or "**confinement time**") of the charged particles in the plasma.

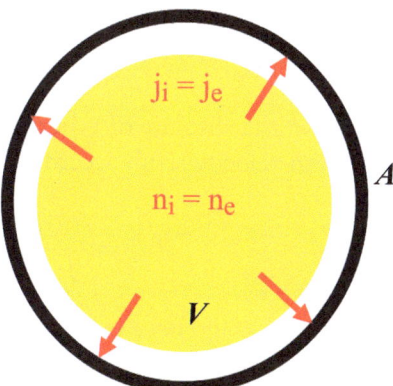

Figure 1.1: A quasineutral plasma with positive ions (density n_i) and electrons (density n_e) in a container with volume V and wall area A. The generation of charged particles in the plasma is compensated by losses to the wall with equal fluxes j_i and j_e.

As the kinetic energy transfer in elastic collisions is most efficient for particles of equal mass, both the electron and the heavy particle **subensembles** will thermalize rather efficiently, so that temperatures T_e, T_i, and T_n can be defined for the electrons, ions, and neutrals, respectively. In plasma physics, the particle temperatures are mostly expressed by the energy equivalent kT, where k denotes the Boltzmann constant (see Eq. (1.1)):

$$kT = 1 \text{ eV} \cong T = 11{,}604 \text{ K}. \qquad (1.1)$$

In low-pressure plasmas of low volumes, the collisions between electrons and heavy particles during the confinement time are not frequent enough to provide a thermalization of the hot electrons with the neutrals and ions, so that the temperature of the latter remains low and may even stay close to room temperature. Such plasmas with $T_i \approx T_n \ll T_e$ are called "**non-thermal.**" In contrast, "**thermal**" plasmas are characterized by $T_i \approx T_e$.

Further classifications are used with respect to the electron temperature and the pressure of the plasma. The electron temperature of "**low-temperature**" plasmas is in the order of a few eV, "high-temperature" plasmas achieve electron temperatures exceeding 10 keV.

Plasmas are called "**low-pressure**" at pressures below app. 1 mbar or 100 Pa, "**high-pressure**" above app. 100 mbar. Of special relevance for technical application are "**atmospheric**" plasmas at a pressure around 1 bar.

1.2   MAXWELLIAN TEMPERATURE DISTRIBUTION

Assuming a sufficiently number of collisions, the energy distribution of a plasma species can be simplistically described by the Maxwellan distribution function, which maximizes entropy. The distribution over the three-dimensional velocity space is given by

$$f(v) = \left(\frac{m}{2\pi kT}\right)^{\frac{3}{2}} \exp\left(-\frac{mv^2}{2kT}\right), \qquad (1.2)$$

where the distribution depends only on the absolute of v in case of an isotropic plasma. From this, the mean kinetic energy and mean velocity result according to

$$\langle E \rangle = \frac{\int mv^2 f(v) d^3v}{2 \int f(v) d^3v} = \frac{3}{2}kT \qquad \langle v \rangle = \frac{\int v f(v) d^3v}{\int f(v) d^3v} = \sqrt{\frac{8kT}{\pi m}}. \qquad (1.3)$$

(Actually, the integral in the denominator could have been omitted for the case of Eq. (1.2), as this function in normalized to 1.) Transforming into the energy distribution function yields

$$f(E) = 4\pi v^2 f(v) \left|\frac{dv}{dE}\right| = \frac{2}{\sqrt{\pi}} E^{\frac{1}{2}} (kT)^{-\frac{3}{2}} \exp\left(\frac{E}{kT}\right) \qquad (1.4)$$

which is plotted in Fig. 1.2.

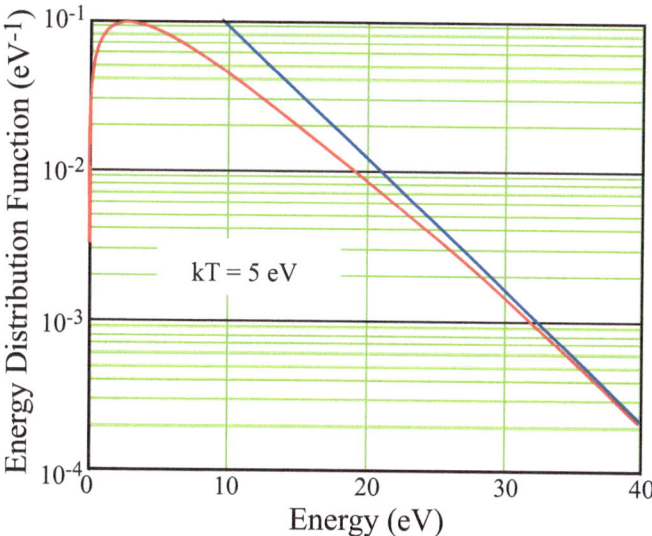

Figure 1.2: Maxwellian energy distribution function for a temperature of 5 eV. In the limit of high energy, the energy dependence converges to $\sim \exp(-E/kT)$ (blue line).

1.3   DEBYE LENGTH

Quasineutrality can be quantified by the Debye length which is the maximum distance in a plasma by which charges can be separated. According to the Poisson equation, a local deviation of the charged particle atomic density from neutrality, δ_n, along a distance λ results in an electrostatic potential difference U with

$$\Delta U \approx \frac{U}{\lambda^2} \approx \frac{e}{\varepsilon_0} \delta n, \tag{1.5}$$

where e denotes the elementary charge. Charge fluctuations are caused by the thermal motion of the electrons, which results via

$$kT_e \approx eU \tag{1.6}$$

in the maximum separation λ. For full charge separation, i.e., $\delta n = n_e$, the Debye length results as

$$\lambda_D = \sqrt{\frac{\varepsilon_0 kT_e}{n_e e^2}}. \tag{1.7}$$

Strong static electrical fields can only exist with the **Debye sphere** of radius λ_D. To fulfill quasineutrality, the linear dimensions of the plasma L have to be large compared to the Debye length

$$L \gg \lambda_D. \tag{1.8}$$

Static collective behavior the plasma requires a sufficiently large number of charged particles within the Debye sphere, i.e.,

$$\frac{4}{3}\pi\lambda_D^3 \cdot n_e \gg 1. \tag{1.9}$$

Via Eq. (1.7), this imposes the condition of a minimum electron temperature at given electron density.

1.4   PLASMA FREQUENCY

Similarly, the request for **dynamic collective behavior** can be quantified by the plasma frequency. The simplest collective motion is the oscillation of the (fast) electrons against the ion background (assumed to be at rest). Displacement of a sheet of electrons (see Fig. 1.3) creates an electric field

$$E = \frac{en_e\delta}{\varepsilon_0}, \tag{1.10}$$

where δ is the displacement and thereby also thickness of the sheet. The equation of motion of the electron sheet is

$$M\frac{d^2\delta}{dt^2} = -QE \tag{1.11}$$

with the mass per unit area $M = n_e\delta m_e$, where m_e denotes the electron mass, and the charge per unit area $Q = n_e\delta e$. Inserting Eq. (1.10), yields an equation of oscillation with the angular frequency

$$\omega_{pe} = \sqrt{\frac{n_e e^2}{\varepsilon_0 m_e}} \tag{1.12}$$

which is called the (electron) plasma frequency.

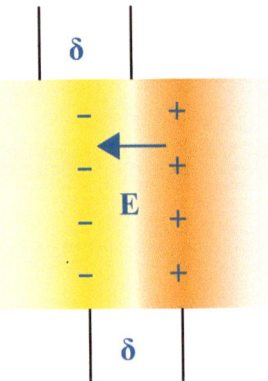

Figure 1.3: Displacement of a sheet of electrons against the ion background.

The collective oscillation of the electrons should not be impeded by collisions with the heavy particles. If their mean collision angular frequency is denoted by ω_c,

$$\omega_{pe} \gg \omega_c \qquad (1.13)$$

can be formulated as a condition for dynamic collective behavior.

1.5 CLASSIFICATION OF PLASMAS

As seen in the previous sections, the main characteristics of the plasma are determined by the electron temperature, T_e, and the electron density, n_e. Correspondingly, different plasma can also be identified by their coordinates in a plot of kT_e vs. n_e. This is shown in Fig. 1.4.

As seen, the electron density of known plasmas may vary over about 30 orders of magnitude, their electron temperature over about 7 orders of magnitude. All shown plasmas can be treated **non-relativistically**, as indicated by the green line.

If the mean kinetic energy of electrons falls below the Coulomb potential of the interaction of electrons at their mean atomic distance,

$$V_C = \frac{e^2}{4\pi \varepsilon_0 \langle r_e \rangle} = \frac{e^2 n_e^{1/3}}{4\pi \varepsilon_0} \qquad (1.14)$$

plasmas are, in accordance to gas theorie, called "**non-ideal**." The red transition line in Fig. 1.4 is given by $\langle E \rangle = V_C$ and scales, according to Eq. (1.3), as $kT_e \sim n_e^{1/3}$.

In "**degenerate**" plasmas, the mean kinetic energy of electrons is smaller than the Fermi energy of a free electron gas, which is given by

$$E_F = \frac{\hbar^2}{2m_e} \left(3\pi^2 n_e \right)^{2/3}. \qquad (1.15)$$

The red line in Fig. 1.4 results from $\langle E \rangle = E_F$ and scales as $kT_e \sim n_e^{2/3}$.

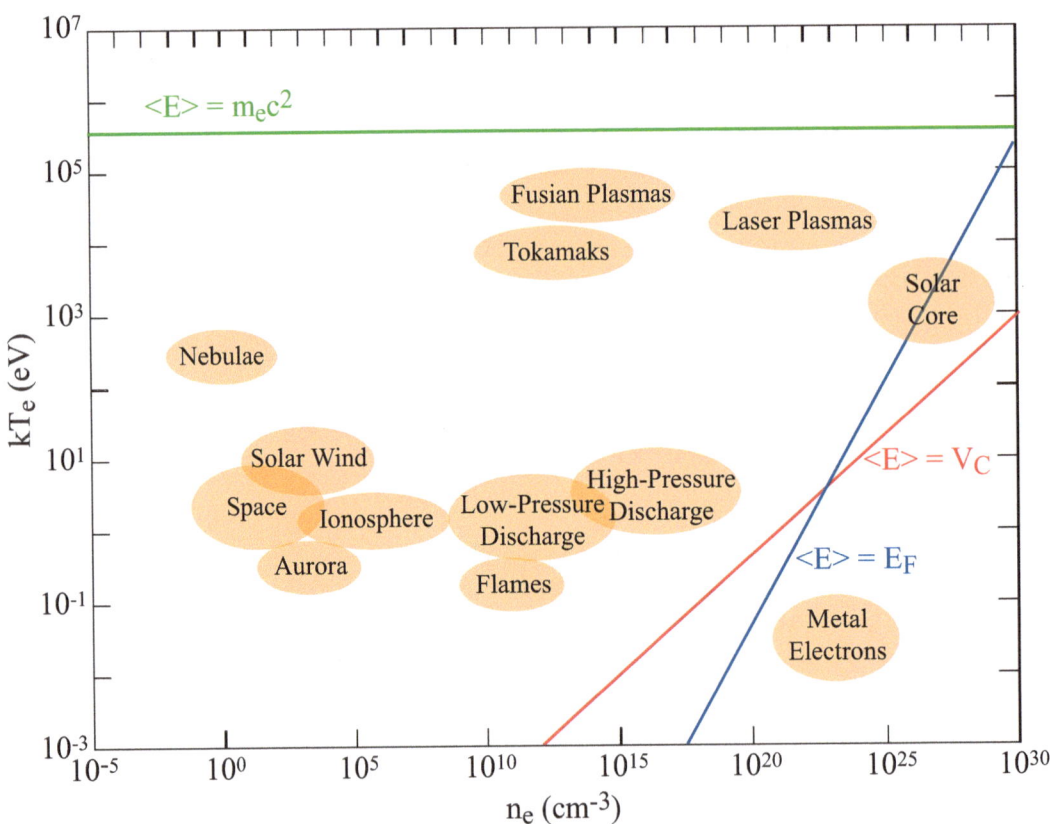

Figure 1.4: Different kind of plasmas in astrophysics, solar physics and solid state physics, nuclear fusion, and for technical applications in a plot of electron temperature vs. electron density. The green line denotes the limitation of nonrelativistic plasmas, degenerate and non-ideal plasmas are positioned right from the blue and red lines, respectively.

CHAPTER 2

Collisions in Plasmas

2.1 GENERAL DEFINITIONS

For a moving particle in an ensemble of identical or different particles with atomic density n, the average number of collision processes in a path length interval Δs is given by

$$\Delta N_c = n\sigma\Delta s, \tag{2.1}$$

where σ denotes the collision cross section (see Fig. 2.1).

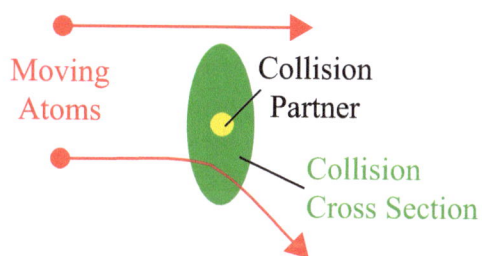

Figure 2.1: Definition of the collisional cross section: A moving particle hitting a collision partner within the area σ is deflected, otherwise not.

One collision in average defines the collisional **mean free path length**

$$\lambda_c = \frac{1}{n\sigma}. \tag{2.2}$$

From this, the **mean collision time** for a particle with velocity v is

$$t_c = \frac{\lambda_c}{v} = \frac{1}{nv\sigma}. \tag{2.3}$$

If the moving particle belongs to an ensemble with a distribution of velocities, the product $v\sigma$ has to be averaged over the velocity distribution, as the cross section generally depends on the velocity. This results in the collisional **"rate coefficient"**

$$\langle \sigma v \rangle = \frac{\int v^3 \sigma(v) f(v) dv}{\int v^2 f(v) dv}. \tag{2.4}$$

Then, the **collision frequency** becomes

$$\nu_c = n \langle \sigma v \rangle. \tag{2.5}$$

2.2   BINARY ELASTIC COLLISION KINEMATICS

In the simplest approximation, the collisions between charged and neutral particles in a plasma can be treated as binary elastic collision. In most plasma, many-body interaction can be neglected due to the large mean free path lengths, as, e.g., compared to atomic collisions in a solid.

The two-body scattering problem for a given spherically symmetric interaction potential $V(r)$ is reduced to a one-body scattering problem by transformation from the laboratory system (LS) into the center-of-mass system (CMS) (see Fig. 2.2).

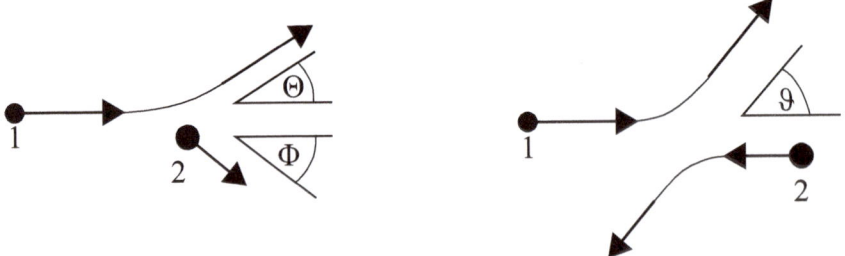

Figure 2.2: Elastic binary collision of a moving particle 1 with a particle 2 being at rest before the collision, in the laboratory system (left) and the center-of-mass system (right).

The resulting equation of motion in the CMS yields single-particle scattering kinematics

$$\mu \frac{d^2 \vec{R}}{dt^2} = -\vec{\nabla} V(R), \qquad \vec{R} = \vec{R}_1 - \vec{R}_2 \tag{2.6}$$

with the **reduced mass** (m_1 and m_2 denote the mass of the projectile and the target, respectively)

$$\mu = \frac{m_1 m_2}{m_1 + m_2}. \tag{2.7}$$

The **energy in the CMS system**, being available for the collision, is given by

$$E_c = \frac{m_2}{m_1 + m_2} E \tag{2.8}$$

with E denoting the projectile energy in the LS. In collisions of electrons with heavy particles LS and CMS are practically identical.

Momentum and energy conservation yield the transformations of the asymptotic scattering angles between CMS and LS for an elastic collision (for definitions see Fig. 2.2)

$$\tan \Theta = \frac{\sin \vartheta}{\frac{m_1}{m_2} + \cos \vartheta} \qquad \Phi = \frac{\pi - \vartheta}{2} \tag{2.9}$$

and the reverse transformation

$$\vartheta = \Theta + \arcsin \left(\frac{m_1}{m_2} \sin \Theta \right). \tag{2.10}$$

The **energy transfer** to the target atom ("recoil") (in LS) is given by

$$T = \gamma E \sin^2 \frac{\vartheta}{2} \tag{2.11}$$

with the energy transfer factor

$$\gamma = \frac{4m_1 m_2}{(m_1 + m_2)^2}. \tag{2.12}$$

From this, the LS projectile energy after the collision becomes

$$E' = E - T = E \left(1 - \frac{4m_1 m_2}{(m_1 + m_2)^2} \sin^2 \frac{\vartheta}{2} \right) \tag{2.13}$$

or after transformation into the LS according to (2.9)

$$E' = E \left(\frac{m_1}{m_1 + m_2} \right)^2 \left(\cos \Theta \pm \sqrt{\left(\frac{m_2}{m_1} \right)^2 - \sin^2 \Theta} \right)^2. \tag{2.14}$$

For collisions at least in a one-species plasma, either $m_1 = m_2$ (electron-electron, neutral-neutral, and ion-neutral collisions) or $m_1 \ll m_2$ (collisions with electrons with ions or neutrals) hold. In these cases, only the positive sign in Eq. (2.14) is valid.

2.3   DIFFERENTIAL CROSS SECTION

As seen from Fig. 2.3, a moving particle entering asymptotically a ring with the radius of the impact parameter p, will be deflected with a corresponding polar deflection angle θ. The partial cross section, i.e., the area of the ring,

$$d\sigma = 2\pi p \, dp \tag{2.15}$$

corresponds to a solid angle of (polar coordinates, axial symmetry)

$$d\omega = 2\pi \sin \vartheta \, d\vartheta. \tag{2.16}$$

Combining Eqs. (2.15) and (2.16) yields the CMS differential cross section

$$\frac{d\sigma}{d\omega} = \frac{p}{\sin \vartheta} \left| \frac{dp}{d\vartheta} \right|. \tag{2.17}$$

Equation (2.17) requires a known relation between θ and p. This is given for classical scattering by the "**classical trajectory integral**," which results from the conservation of energy and angular momentum

$$\vartheta = \pi - 2p \int_0^{R_{\min}^{-1}} \frac{d\left(\frac{1}{R}\right)}{\sqrt{1 - \frac{V(R)}{E_c} - \frac{p^2}{R^2}}}. \tag{2.18}$$

In Eq. (2.18), $R_{\min}$ denotes the distance of minimum approach and is obtained by setting the denominator of the integrand to 0.

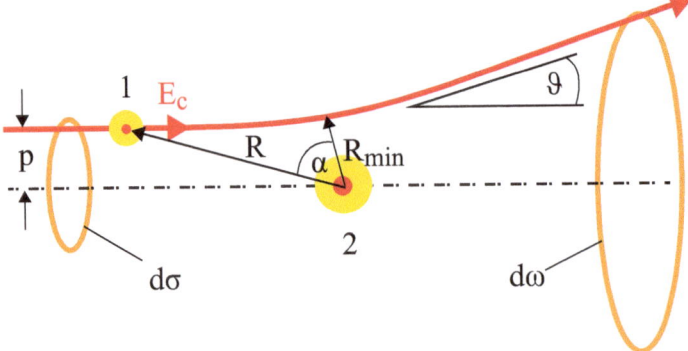

Figure 2.3: Trajectory of a scattered particle 1 with the impact parameter p and the deflection angle θ.

2.4   MOMENTUM TRANSFER

For transport calculations, the probability of scattering might be less interesting than the transfer of momentum during the collision. After the collision with initial momentum p_0, the remaining momentum in the original direction of motion is (see Fig. 2.4)

$$p_0 - \Delta p_0 = p_0 \cos \vartheta. \tag{2.19}$$

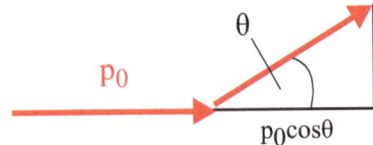

Figure 2.4: Momentum transfer. Note that the length of the momentum vector remains constant after elastic scattering in the CMS.

So, the relative momentum transfer is

$$\frac{\Delta p_0}{p_0} = 1 - \cos \vartheta \tag{2.20}$$

and the differential **momentum transfer cross section** becomes

$$\frac{d\sigma_{MT}}{d\omega} = \frac{d\sigma}{d\omega}(1 - \cos \vartheta). \tag{2.21}$$

For the example of a Coulomb collision, forward scattering is strongly preferred, but is associated with low momentum transfer. This is accounted for by the correction of the cross section. The resulting total momentum cross section becomes

$$\sigma_{MT} = 2\pi \int_{0}^{\pi} \frac{d\sigma}{d\omega}(1 - \cos\vartheta)\sin\vartheta\,d\vartheta. \tag{2.22}$$

2.5   COULOMB COLLISIONS

For two point charges q_1 and q_2 at distance R, the Coulomb interaction potential reads

$$V_c(R) = \frac{q_1 q_2}{4\pi\varepsilon_0 R}. \tag{2.23}$$

Inserting this into Eq. (2.18) and integrating yields

$$\tan\frac{\vartheta}{2} = \frac{b}{2p} \tag{2.24}$$

with the "**collision diameter**" (the minimum distance of approach at a central collision of two charges of equal sign—see Eq. (2.23))

$$b = \frac{q_1 q_2}{4\pi\varepsilon_0 E_c}. \tag{2.25}$$

According to Eq. (2.17), the **Rutherford** differential scattering cross section then results as

$$\frac{d\sigma}{d\omega} = \left(\frac{b}{4\sin^2(\vartheta/2)}\right)^2. \tag{2.26}$$

Thus, the Rutherford cross section scales as $1/E^2$ and strongly favors forward collisions.

The total cross section, as obtained by integration over the solid angle, diverges due to the infinite range of the Coulomb potential. (However, it should be noted that the momentum transfer cross section, Eq. (2.22), is finite.) Therefore, for an estimation of a total cross section for "significant deflections" one may use the cross section for all events with a scattering angle exceeding 90° (all backscattering events in the CMS) as an estimate, which results as

$$\sigma\left(90°\right) = \frac{\pi}{4}b^2. \tag{2.27}$$

Due to the strong forward characteristics of Rutherford scattering, most of the collisions included in Eq. (2.27) take place with scattering angles closely above 90°.

Alternatively to this single scattering event, all scattering angles can also be reached by a sequence of low-angle scattering events ("multiple scattering") (see Fig. 2.5).

Equation (2.24) reads for small scattering angles

$$\vartheta = \frac{b}{p}. \tag{2.28}$$

Figure 2.5: A scattering angle of 90° by a single collision (left) or a sequence of small angle collisions (right).

Simultaneously, the CMS-LS transformation (Eq. (2.9)) becomes

$$\Theta = \frac{m_2}{m_1 + m_2} \vartheta. \tag{2.29}$$

Combining Eqs. (2.8), (2.25), (2.28), and (2.29) demonstrates that Eq. (2.28) can now be applied to the laboratory system, i.e.,

$$\Theta = \frac{B}{p}, \tag{2.30}$$

where B is the collision diameter defined with the laboratory energy. Therefore, consecutive small-angle scattering in the LS can be treated without transformation.

As the mean deflection angle is 0 due to cylindrical symmetry, the variance of the scattering angle for a single scattering event is given by

$$\langle(\Theta - \langle\Theta\rangle)^2\rangle = \langle\Theta^2\rangle = \frac{1}{\pi p_{max}^2} \int_{p_{min}}^{p_{max}} \frac{B^2}{p^2} 2\pi p \, dp = 2\frac{B^2}{p_{max}^2} \log\frac{p_{max}}{p_{min}} \tag{2.31}$$

as $p_{min} \ll p_{max}$. The maximum impact parameter is identified with the Debye length, as charges become completely screened outside the Debye sphere. With $p_{min} = B$, the logarithm in Eq. (2.31) becomes the so-called **Coulomb logarithm**

$$\log \Lambda = \log\frac{\lambda_D}{B}. \tag{2.32}$$

For N_c collisions within a path length s, the individual variances add, resulting in a total variance according to Eq. (2.1)

$$\langle(\Theta(N_c))^2\rangle = N_c \langle\Theta^2\rangle = \langle\Theta^2\rangle \pi p_{max}^2 n s \tag{2.33}$$

as the total cross section of the single collision is πp_{max^2}. Alternatively, the multiple scattering process can also be regarded as one scattering event. Again according to Eq. (2.1), its cross section is

$$\sigma_{ms} = \frac{1}{ns} \qquad (2.34)$$

resulting in

$$\frac{\sigma_{ms}}{\pi p_{max}^2} = \frac{\langle \Theta^2 \rangle}{\langle (\Theta (N_c))^2 \rangle}. \qquad (2.35)$$

This we evaluate again for 90° multiple scattering with

$$\left\langle (\Theta (N_c))^2 \right\rangle = \left(\frac{\pi}{2} \right)^2 . \qquad (2.36)$$

Combining Eqs. (2.31), (2.32), (2.35), and (2.36) yields

$$\sigma_{ms} (90°) = \frac{8}{\pi} B^2 \log \Lambda. \qquad (2.37)$$

Comparing with Eq. (2.27) yields the ratio of the multiple scattering to single scattering cross sections

$$\frac{\sigma_{ms}(90°)}{\sigma(90°)} = \frac{32}{\pi^2} \left(\frac{m_2}{m_1 + m_2} \right)^2 \log \Lambda. \qquad (2.38)$$

$\log \Lambda \approx 10$ is a good approximation for the Coulomb logarithm in many plasmas. Thereby, the Coulomb scattering in a plasma is largely determined by multiple scattering rather than single scattering.

2.6 COLLISIONS OF NEUTRALS

Neutral particles in a plasma are in general not only represented by the non-ionized fraction of the parent gas, but may also result from plasma-chemical processes such as the electron-induced dissociation and chemical reaction in the plasma. As they are cold in most plasmas, any energy transfers as, e.g., by electronic excitations, can normally be neglected. Then, for their collisions a hard sphere approximation is appropriate which is sketched in Fig. 2.6.

The cross section is then given by

$$\sigma = \pi (r_A + r_B)^2 , \qquad (2.39)$$

where r_A and r_B denote the atomic radii of the collision partners. These cross sections are typically in the range $(2 \ldots 5) \cdot 10^{-15} \, cm^2$.

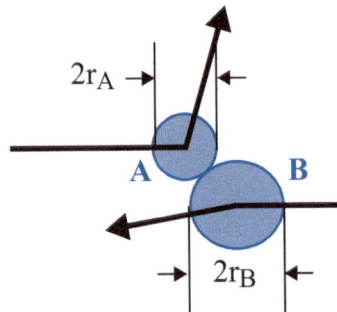

Figure 2.6: Hard-sphere collisions of two A and B with the respective atomic radii.

2.7   RESONANT CHARGE TRANSFER

When an ion collides with a neutral, a variety of processes may occur, such as the transfer of electronic excitation, in particular when the ion is fast compared to the neutral. This is the case in thermal plasmas, but also, as will be shown later, in the boundary layer of low-pressure, non-thermal plasma where the ion temperature in the plasma is low. A prominent process with an often very large cross section is the charge transfer between an ion and a neutral. Assuming a fast ion and a thermal neutral with negligible velocity, the process is at a small impact parameter

$$A^+(fast) + B(at\ rest) \Rightarrow A(fast) + B^+(at\ rest). \tag{2.40}$$

This is a two-step process, the release of the electron from B and the capture by A^+. At a center-to-center separation a_{AB}, the potential energy of a B electron in a level n is in a simplified atomic picture

$$W_{nB} = -\frac{U_{iB}}{n^2} - \frac{e^2}{4\pi\varepsilon_0 a_{AB}}, \tag{2.41}$$

where U_{iB} is the ionization energy of B in level $n = 1$. The second term describes the interaction with the ion. After release from B, the electron obtains the potential energy (see Fig. 2.7)

$$U(z) = -\frac{e^2}{4\pi\varepsilon_0 z} - \frac{e^2}{4\pi\varepsilon_0 |a_{AB} - z|}. \tag{2.42}$$

From this, the maximum of $U(z)$ between A and B results as

$$U_{max} = -\frac{e^2}{\pi\varepsilon_0 a_{AB}}, \tag{2.43}$$

i.e., the potential barrier decreases when the collision partner approach. The classical condition for release of the electron from B is given by $W_{nB} > U_{max}$, which, by equating Eqs. (2.41) and (2.43), results in a maximum distance at which the release takes place

$$a_{AB,r} = -\frac{3e^2 n^2}{4\pi\varepsilon_0 U_{iB}}. \tag{2.44}$$

When the ion velocity is low compared to the orbital velocities of the electrons, which is the case for non-thermal plasmas, there is sufficient time for the transfer of the released electron to the ion, provided there is a free level below the releasing level of B. Then, the cross section may be estimated from the release only, resulting in

$$\sigma_{COB} = \pi a_{AB,r}^2 = 9\pi \left(\frac{e^2 n^2}{4\pi \varepsilon_0 U_{iB}} \right)^2. \tag{2.45}$$

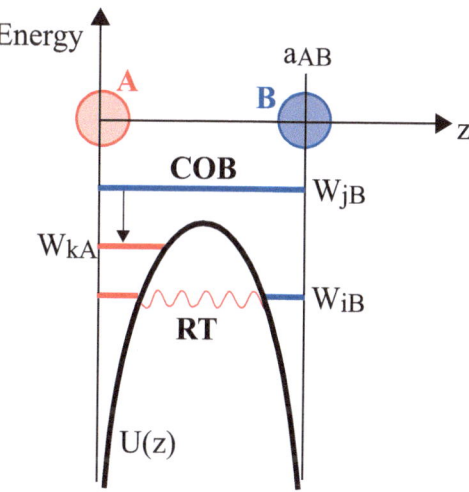

Figure 2.7: Charge transfer from B to A in the classical over-the-barrier model (COB) and due to resonant tunneling (RT).

This is the so-called **classical "over-the-barrier" model**. A further possibility is the resonant charge transfer by tunneling between identical species, i.e., $A = B$. Without explicit treatment here, the cross section can be approximated by

$$\frac{\sigma_{RT}}{cm^2} = \frac{1}{U_i/eV} \left(C_1 - C_2 \log \left(\frac{v_+}{cm/s} \right) \right)^2 \tag{2.46}$$

with U_i denoting the ionization potential and constants $C_1 = 1.58 \cdot 10^{-7}$ and $C_2 = 7.24 \cdot 10^{-8}$. As the tunneling probability increases with the interaction time, the cross section decreases with the ion velocity v_+.

Resonant tunneling dominates the charge exchange in atomic gases. This process,

$$A(slow) + A^+(fast) \Rightarrow A(fast) + A^+(slow) \tag{2.47}$$

may affect the ion transport in weakly ionized plasmas significantly. In particular, it generates fast neutrals.

It should be noted that, in the case of molecules, the probability of resonant charge transfer is low, due to the many open channels of vibrational and rotational excitation of the molecule which violate the condition of resonant energy levels.

2.8   POLARIZATION SCATTERING

Polarization scattering is related to the scattering of a charged particle at a neutral gas atom, the electron cloud of which polarizes when the charged particle is approaching (see Fig. 2.8).

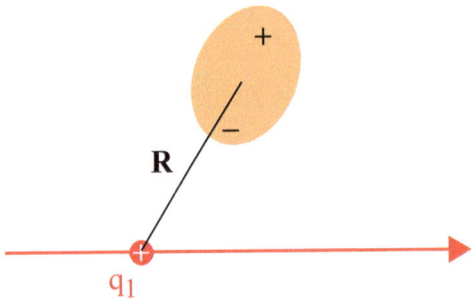

Figure 2.8: Polarization of a neutral atom during the passage of a charged particle.

The electric field of the charged particle is

$$\vec{E} = \frac{q_1}{4\pi\varepsilon_0 R^2}\frac{\vec{R}}{R}. \tag{2.48}$$

With the polarizability α, the induced dipole moment of the atom becomes

$$\vec{p} = \alpha\vec{E} \tag{2.49}$$

and the potential energy

$$V = -\vec{p}\cdot\vec{E} = \alpha E^2 = \frac{q_1^2}{(4\pi\varepsilon_0)^2}\frac{\alpha}{R^4}. \tag{2.50}$$

The strong radial dependence of this potential results below a certain impact parameter p_L in a trajectory which spirals into the neutral, i.e., the charged particle is captured. This situation is sketched in Fig. 2.9.

Calculations show for the corresponding "**Langevin capture cross section**"

$$\sigma_L = \pi p_L^2 = \sqrt{\frac{\pi\alpha q_1^2}{\varepsilon_0\mu}\frac{1}{v_r}}, \tag{2.51}$$

where v_r is the relative velocity of the collision partners before the collision.

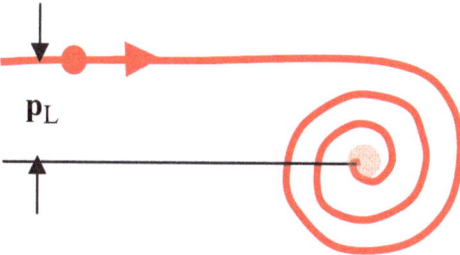

Figure 2.9: Langevin capture of a charged particle due to target polarization.

The Langevin capture of electrons is a mechanism for the formation of negative ions by electron attachment in gases with high electron affinity. It should be noted that the electron affinity does not enter Eq. (2.51), but naturally electron capture only occurs in case of positive electron affinity. As the electron affinity, however, enters the threshold energy for re-neutralization of the negative ion in analogy to electron-induced ionization (see Section 2.10), the concentration of negative ions, which results from the balance of Langevin electron capture and re-neutralization, increases at increasing electron affinity.

2.9 ELECTRON ELASTIC SCATTERING AT NEUTRALS

The treatment of the elastic scattering of electrons at neutral heavy particles and its energy dependence is difficult and would require complex quantum-mechanical calculations. For low energies, the cross section steeply decreases with increasing velocity, as it is mainly given by the interaction time. After a pronounced minimum at an electron energy of some 0.1 eV, it becomes dominated by the collisions in the screened Coulomb potential of the atomic nuclei, which merge into Coulomb collisions in the limit of high energy. An example will be given below (see Fig. 2.13).

2.10 ELECTRON IMPACT IONIZATION

In principal, the ionization of an atom by an electronic collision

$$e + A \Rightarrow A^+ + 2e \tag{2.52}$$

requires a quantum-mechanical treatment. Here we present a simple classical approximation. Applying trigonometric relations, the energy transfer (see Eq. (2.11)) for a collision of equal masses (two electrons) is given by

$$T + \frac{E}{2}(1 - \cos\vartheta). \tag{2.53}$$

As small angles dominate the scattering, the cosine is expanded as

$$\cos \vartheta \approx 1 - \frac{\vartheta^2}{2}. \tag{2.54}$$

According to Eq. (2.10),

$$\vartheta = 2\Theta \tag{2.55}$$

so that

$$T = E\Theta^2. \tag{2.56}$$

The transformation of the Rutherford cross section (Eq. (2.26)) into the LS yields

$$\frac{d\sigma}{d\Omega} = \left(\frac{e^2}{4\pi\varepsilon_0}\right)^2 \frac{1}{E^2} \frac{1}{\Theta^4} \tag{2.57}$$

which is transformed using Eq. (2.56) into

$$\frac{d\sigma}{dT} = \pi \left(\frac{e^2}{4\pi\varepsilon_0}\right)^2 \frac{1}{E} \frac{1}{T^2}. \tag{2.58}$$

Integration from the ionization potential U_i as the minimum necessary energy transfer for ionization, to the maximum energy transfer E yields the **Thomson cross section of ionization**

$$\sigma_i = \pi \left(\frac{e^2}{4\pi\varepsilon_0}\right)^2 \frac{1}{E} \left(\frac{1}{U_i} - \frac{1}{E}\right). \tag{2.59}$$

For practical application, however, this classical formula in the small-angle approximation is mostly of little use. Figure 2.10 compares the result for the electron-induced ionization of Argon to the corresponding experimental data. As seen, Eq. (2.59) largely underestimates the real data over most of the energy range. The maximum of the cross section results at a too low energy. Only in the very threshold regime is there reasonable agreement.

Figure 2.11 displays again an ionization cross section as calculated by the Thomson formula. Related to it is the ionization rate coefficient for a Maxwellian velocity distribution of the electrons, which has been calculated by numerical integration using Eqs. (1.2) and (2.4).

The rate coefficient cannot be obtained analytically due to the occurrence of an $\exp(-z)/z$ integrand. However, the governing energy dependence in the threshold regime, where $kT_e < U_i$, is given by an exponential term of the form

$$\langle \sigma_i v \rangle = Const \cdot \exp\left(-\frac{U_i}{kT_e}\right). \tag{2.60}$$

This temperature dependence is also shown as a fitted curve in Fig. 2.11. It results from the fact that in the limit of low electron temperature, the rate coefficient is determined by the exponential tail of the Maxwell distribution.

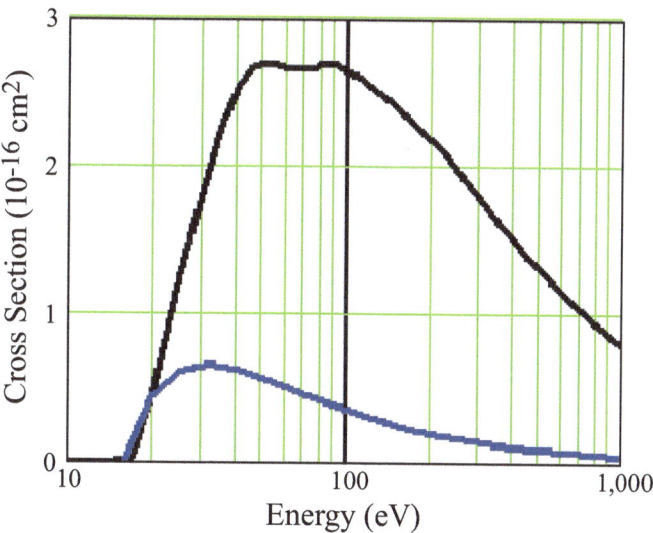

Figure 2.10: Cross section of the electron impact ionization of Ar according to the Thomson formula (blue line) and from experimental data (black line). The ionization potential of Ar is 15.76 eV.

In molecular gases, the ionization might be accompanied by the dissociation of the molecule, provided the electron energy is sufficiently high. For diatomic molecules,

$$e + AB \Rightarrow A^+ + B + 2e. \tag{2.61}$$

The energy dependence of the **dissociative ionization** is similar to that of the atomic ionization, so that Eq. (2.59) holds with a different threshold energy of dissociative ionization. The threshold energy is generally not the sum of the ionization and dissociation thresholds, as ionization and dissociation due not occur independently, and their interaction depends of the electronic energy levels of the individual molecule.

2.11   ELECTRON IMPACT DISSOCIATION

Also, the molecular dissociation by electron impact

$$e + AB \Rightarrow A + B + 2e \tag{2.62}$$

can be treated in a similar way as the electron-impact dissociation, with the knock-on electron now being excited to an antibinding state or to a higher energy level of the molecule which allows auto-dissociation. Correspondingly, the Thomson formula is applied with the dissociation energy of the molecule as the threshold energy.

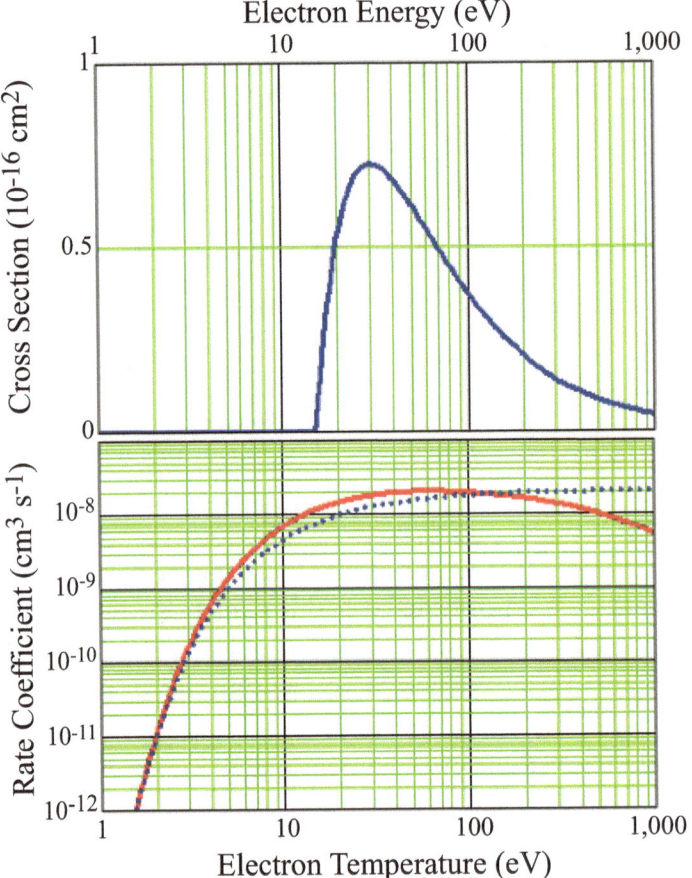

Figure 2.11: Electron impact ionization cross section from the Thomson formula (top) and Maxwell-averaged ionization rate coefficient (bottom) for an ionization potential of 15 eV. The red line results from numerical integration, the blue dotted line represents the low-temperature fit function.

Figure 2.12 demonstrates the pronounced similarity of the cross sections, and thereby the rate coefficients, of ionization, dissociation, and dissociative ionization at the example of the methane molecule. There is only a weak shift of the curves which respect to each other due to the different threshold energies. According to Eq. (2.59), the cross section is highest for the lowest threshold energies.

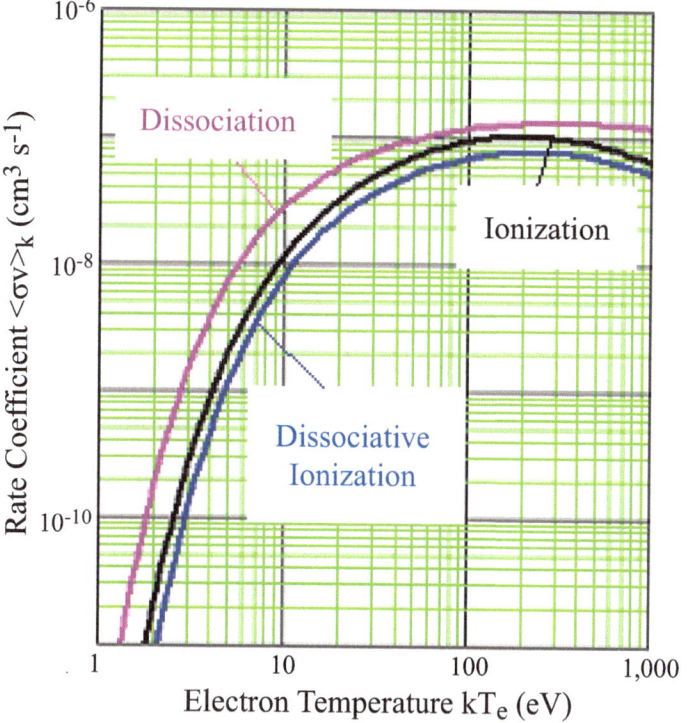

Figure 2.12: Rate coefficients of the electron impact ionization, dissociation, and dissociative ionization of methane. The processes and threshold energies are given in the table.

2.12 ELECTRON IMPACT EXCITATION

The excitation of to higher atomic or molecular electronic levels,

$$e + A \Rightarrow A^* + e \qquad (2.63)$$

or to vibrational levels in molecules can again be treated in a similar way, where the threshold energy is now the excitation energy.

Figure 2.13 compares the energy dependence of the electron impact cross sections of elastic scattering, excitation and ionization for the case of argon. The momentum balance of electrons is largely dominated by the elastic scattering. Excitation and ionization follow a similar energy dependence, with a different threshold and a different magnitude. The magnitiude does not scale with the threshold energy as given in the Thomson formula.

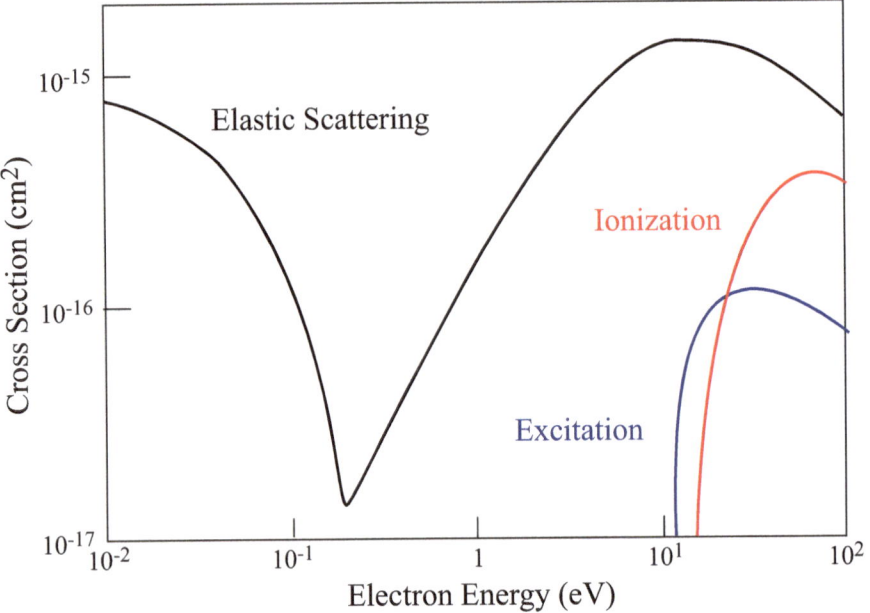

Figure 2.13: Cross sections for electron elastic scattering and electron-induced excitation and ionization in Argon. The threshold energies for ionization and excitation are 15.76 eV and 11.55 eV, respectively.

2.13   PENNING IONIZATION

The Penning process

$$A + B^* \Rightarrow A^+ + B + e \qquad (2.64)$$

can be a significant mechanism of ionization when highly excited metastable atoms are present in the plasma. It requires more than one neutral species as the excitation energy of the excited atom has to exceed the ionization energy of the collision partner. The most efficient metastable atom is He with an excitation energy of 19.82 eV and 20.6 eV for the 2^3 S and the 2^1 S state, respectively (ionization energy 24.5 eV). This is largely sufficient to ionize, e.g., Argon with an ionization energy of 15.76 eV. The cross section reaches in this case a value of 10^{-15} cm^{-2}.

2.14   CHEMICAL REACTIONS

In a plasma with reactive species and sufficiently high pressure, chemical reactions occur, like exothermic neutral bond rearrangements as, e.g.,

$$AB + CD \Rightarrow AC + BD. \qquad (2.65)$$

The rate coefficients are generally rather small and in the order of 10^{-11} cm^3/s. If ions take part, as, e.g.,

$$AB + B^+ \Rightarrow AB^+ + B \tag{2.66}$$

the cross sections become larger and close to the Langevin cross section of Section 2.7. An example is the hydrogen abstraction reaction, which becomes important for mixed argon- hydrogen plasmas in the form $H_2^+ + Ar \Rightarrow ArH^+ + H^0$. As a further example, the reaction $O^+ + O_3 \Rightarrow O_2^+ + O_2$ has a rate coefficient of $\sim 10^{-10}$ cm^3/s.

A large variety of plasma-chemical reaction is employed in industrial plasma processing. In most cases, the elementary collision data are not known. Even with known thermochemical data, their application is very uncertain, as the reactions might be enhanced by electron- induced excitation of the reaction partners. This is the principle of "cold" processing in low-temperature reactive plasmas, but prevents often from reliable plasma modeling.

CHAPTER 3

Motion of Charged Particles

3.1 EQUATION OF MOTION

The motion of a particle with mass m and charge q in electric and magnetic fields obeys the **Lorentz equation**

$$m\frac{d\vec{v}}{dt} = q\left(\vec{E} + \vec{v} \times \vec{B}\right).$$ (3.1)

3.2 CONSTANT MAGNETIC FIELD

In a Cartesian coordinate system with a zero electric field and a constant magnetic field in z direction the axial force in z direction vanishes. The azimuthal equations of motions are

$$m\frac{dv_x}{dt} = qv_y B \quad m\frac{dv_y}{dt} = -qv_x B \quad m\frac{dv_z}{dt} = 0.$$ (3.2)

This corresponds to a circular motion around the z direction with

$$x = r_L \cos\omega_C t \qquad y = \mp r_L \sin\omega_C t$$ (3.3)

in the (x, y) plane (see Fig. 3.1) with the **cyclotron frequency**

$$\omega_C = \frac{|q|}{m} B$$ (3.4)

and the **Larmor radius** which is given by the initial velocity component $v_\perp$ being perpendicular to the magnetic field

$$r_L = \frac{v_\perp}{\omega_C} = \frac{mv_\perp}{|q|B}$$ (3.5)

and a constant velocity in z direction. In Eq. (3.3), the upper and lower signs stand for positive and negative charges, respectively.

Thus, the particle spirals with constant parallel velocity around the magnetic field vector, as shown in Fig. 3.2.

3.3 CONSTANT ELECTRIC AND MAGNETIC FIELDS

For a combination of constant electric and magnetic fields we place the electric field in the (x, z) planes and, as before, B along the z axis. Then,

$$\frac{dv_x}{dt} = \frac{q}{m}E_x \pm \omega_c v_y \qquad \frac{dv_y}{dt} = \mp\omega_c v_x \qquad \frac{dv_z}{dt} = \frac{q}{m}E_z.$$ (3.6)

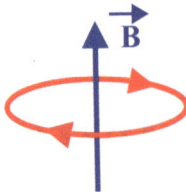

Figure 3.1: Motion of gyration of a charged particle in the (x, y) plane. For a positive charge, the rotation is clockwise with respect to the direction of the magnetic field.

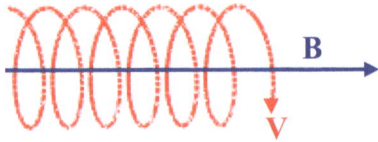

Figure 3.2: Motion of gyration of a charged particle with velocity v around a magnetic field B.

The solution is

$$x = r_L \cos \omega_C t \quad y = \mp r_L \sin \omega_C t - \frac{E_x}{B_z} t \quad z = \frac{q}{2m} E_z t^2 + v_\parallel^0 t. \tag{3.7}$$

where $v_\parallel^0$ denotes the initial velocity component parallel to the magnetic field. This represents the spiral motion with a superimposed drift in the y direction, i.e., perpendicular to the E and B vectors. At $E_z = 0$, i.e., perpendicular electric and magnetic fields, there is again a constant velocity in z direction. This situation is displayed in Fig. 3.3.

We now generalize the Lorentz equation and replace the force exerted by the electric field by any force F which acts in addition to the magnetic field. In accordance with Eq. (3.7), the particle velocity is separated into the time-dependent velocity under the influence of the force and the magnetic field, and a **constant drift velocity** v_D which is perpendicular to the magnetic field. For the latter according to Eq. (3.1),

$$0 = \left(\frac{\vec{F}}{q} + \vec{v}_D \times \vec{B} \right). \tag{3.8}$$

Vector multiplication with B yields

$$0 = \vec{F} \times \vec{B} + q \left(\vec{v}_D \times \vec{B} \right) \times \vec{B} = \vec{F} \times \vec{B} - q \vec{v}_D B^2 + q \vec{B} \left(\vec{B} \cdot \vec{v}_D \right). \tag{3.9}$$

As $v_D \perp B$, the last term vanishes, and the drift velocity becomes

$$\vec{v}_D = \frac{\vec{F} \times \vec{B}}{q B^2}. \tag{3.10}$$

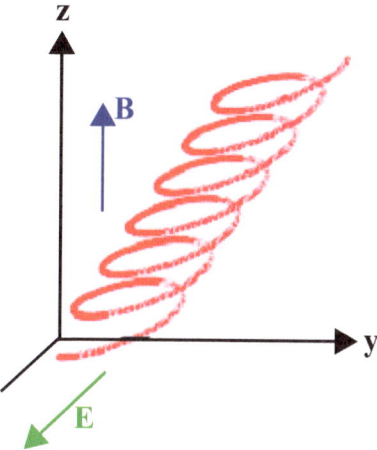

Figure 3.3: Motion of a charged particle in perpendicular electric (E) and magnetic (B) fields.

In the present case of the drift in an electric and a magnetic field, which are both constant, the so-called "**ExB**" **drift** velocity results as

$$\vec{v}_{E\times B} = \frac{\vec{E} \times \vec{B}}{B^2} \tag{3.11}$$

which is in accordance to Eq. (3.7). The drift is directed perpendicular both to the electric and the magnetic field. The guiding center of the drifting gyromotion then moves according to

$$\vec{v}_{GC} = \vec{v}_= + \vec{v}_D, \tag{3.12}$$

where $v_=$ denotes the velocity parallel to the magnetic field, the time dependence of which is governed by the parallel component of the electric field.

3.4   INHOMOGENEOUS MAGNETIC FIELD

For the following, it is assumed that the variation of the magnetic field shall be small along the distance of a gyroradius, i.e.,

$$r_L^{-1} >> \frac{\left|\vec{\nabla} B\right|}{B}. \tag{3.13}$$

First, a **curved magnetic field** is considered. In this case, the centrifugal force acts on the particle, which is given by

$$\vec{F}_c = \frac{m v_=^2}{R_c} \frac{\vec{R}_c}{R_c}, \tag{3.14}$$

where R_c is the radius of curvature. Consequently, the sidewards drift velocity is

$$\vec{v}_C = \frac{m}{q} \frac{v_\parallel^2}{R_c^2} \cdot \frac{\vec{R}_c \times \vec{B}}{B^2}.$$ (3.15)

The geometry is indicated in Fig. 3.4. Unlike the ExB drift, the centrifugal drift separates charges of different polarity. Note that the curvature radius vector points inward in case of a convex magnetic field geometry and outward for a concave one.

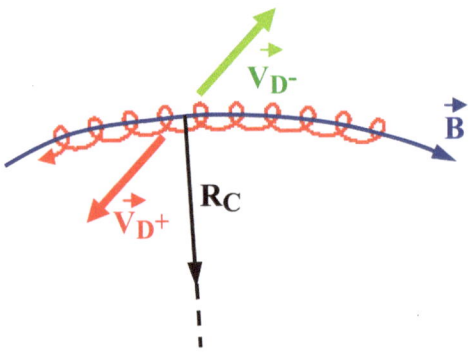

Figure 3.4: Drift motion of a charged particles in a curved magnetic field for a positive charge. Charges of different polarities are separated.

In the following, we will discuss the effect of a **spatial variation of the magnetic field strength**. The gyration in the radial plane than carries a magnetic momentum μ, which relates the magnetic field to the energy of the gyration, $W_\perp$, according to

$$W_\perp = \frac{m}{2} v_\perp^2 = \mu B.$$ (3.16)

The force acting on a magnetic moment (and thereby here on the gyrocenter of the rotating charge) is

$$\vec{F} = -\mu \vec{\nabla} B.$$ (3.17)

If the gradient of the magnetic field strength is perpendicular to the magnetic field, the drift velocity results directly from Eq. (3.10)

$$\vec{v}_{B \times \nabla B} = \frac{m v_\perp^2}{2q} \frac{\vec{B} \times \vec{\nabla} B}{B^3}$$ (3.18)

and, inserting Eq. (3.4),

$$\vec{v}_{B \times \nabla B} = \pm \frac{v_\perp^2}{2\omega_C} \frac{\vec{B} \times \vec{\nabla} B}{B^2}.$$ (3.19)

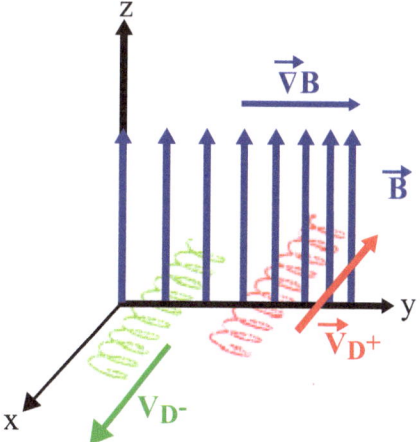

Figure 3.5: Drift motion of a charged particles in an inhomogeneous magnetic field with $B \perp \nabla B$. Charges of different polarities are separated.

This situation is displayed in Fig. 3.5. Again, the $B \times \nabla B$ drift changes direction for different polarities of the charged particles, and this tends to separate positive and negative charge carriers.

If the gradient of B is in the direction of B, as shown in Fig. 3.6 for an axially symmetric geometry, the $B \times \nabla B$ drift vanishes. However, there is an axial force according to Eq. (3.17) which acts on the gyrocenter. In Fig. 3.6, there is a positive gradient of B in z direction so that the force retards any spiral motion toward the region of higher field, independent of the polarity. The axial equation of motion is

$$m\frac{dv}{dt} = -\mu\frac{\partial B}{\partial z}.$$

(3.20)

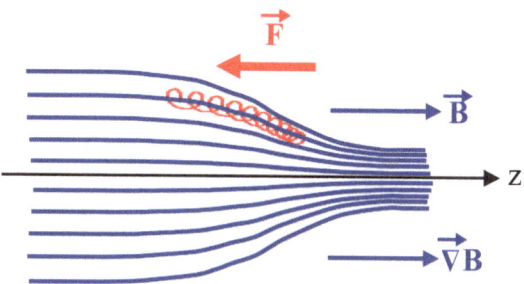

Figure 3.6: Motion of a charged particles in an inhomogeneous magnetic field with $B \parallel \nabla B$. The particle is retarded toward the region of high field.

With $dz = v_= dt$, the variation of the translational kinetic energy results as

$$\frac{dW_=}{dt} = \frac{d}{dt}\left(\frac{m}{2}v_=^2\right) = -\mu\frac{\partial B}{\partial t}. \tag{3.21}$$

Without any additional potential, the total kinetic energy

$$W = W_= + W_\perp \tag{3.22}$$

is conserved. With Eqs. (3.16) and (3.21), this results in

$$-\mu\frac{\partial B}{\partial t} + \frac{\partial}{\partial t}(\mu B) = B\frac{d\mu}{dt} = 0. \tag{3.23}$$

Thus, the **magnetic moment remains constant** during the gyromotion along the inhomogeneous magnetic field. Consequently, transport from a low-field region to a high-field region increases the transversal kinetic energy, and correspondingly decreases the translational kinetic energy, and vice versa. Thus, an inhomogeneous field configuration with $B \parallel \nabla B$ can be employed to accelerate or decelerate charged particles.

Looking at the magnetic flux Φ through the gyro-orbit, we find

$$\frac{d\Phi}{dt} = \frac{d}{dt}\left(\pi r_L^2 B\right) = \frac{\pi m}{q^2}\frac{d}{dt}\left(\frac{mv_\perp^2}{B}\right) = 2\frac{\pi m}{q^2}\frac{d\mu}{dt} = 0 \tag{3.24}$$

which means that the **magnetic flux through the gyro-orbit is conserved** during the transport in the magnetic field.

3.5   GRAVITATION AND MAGNETIC FIELD

According to Eq. (3.10), also gravitation causes a drift when the magnetic field has a horizontal component. With the gravitational acceleration g, the result is

$$\vec{v}_G = \frac{m}{q}\frac{\vec{g} \times \vec{B}}{B^2} \tag{3.25}$$

which again tends to separate positive and negative charges.

3.6   DRIFTS AND INSTABILITIES

As stated above, some of the drifts (gravitational, $B \times \nabla B$) separate charges of different polarity. These may cause instabilities at plasma edges, as it will be discussed for the case of the curvature drift in a plasma geometry shown in Fig. 3.7, the so-called "magnetic mirror" geometry. As seen from Eq. (3.15), the curvature drift acts primarily on the ions. Even if the ion temperature is small compared to the electron temperature, as, e.g., in a low-pressure non-thermal plasma with

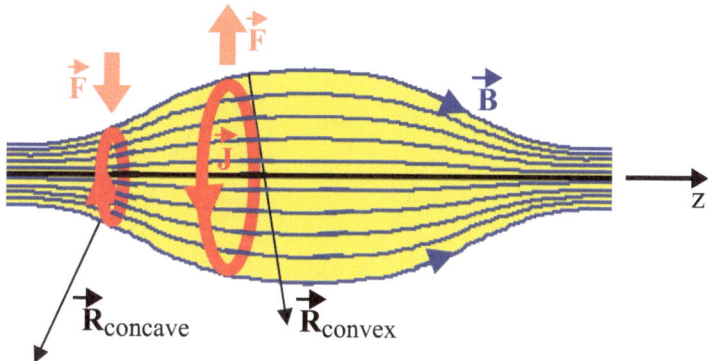

Figure 3.7: Induced drift currents in an axisymmetric "magnetic mirror" configuration, with the Lorenz forces F acting on the plasma at convex and concave curvatures.

$kT_i \approx 0.05$ eV and $kT_e \approx 5$ eV, the parallel velocities of ions and electrons, scaling with the square root of the temperature, differ much less than the masses.

Therefore, a net azimuthal current density $j+$ is induced in the plasma as indicated in the figure ($v_{C,i}$ ion drift velocity)

$$\vec{j}_+ \approx n_e \vec{v}_{C,i}. \tag{3.26}$$

In the case of a convex plasma boundary (waist of the plasma in the middle of Fig. 3.7), the ring current flows clockwise with respect to the magnetic field direction, at a concave curvature counterclockwise.

This results in a Lorentz force per unit volume

$$\vec{f} = e \left[\vec{j}_+ \times \vec{B} \right] \tag{3.27}$$

which is directed outward at convex curvature and inward at concave curvature. Thus, the plasma tends to become unstable in case of convex curvature.

Also, the gravitational drift acts preferentially on the ions. Figure 3.8 shows a situation in which the plasma boundary undergoes fluctuations (e.g., due to electromagnetic waves or statistically) perpendicular to the magnetic field. The gravitational drift pushes the ions to the left for the field geometry of Fig. 3.8, resulting in an ion accumulation at the edges of the instability and an electric field which is directed to the right in the plasma and to the left outside the plasma. The resulting $E \times B$ drift amplifies the instability by pulling the extrusions outside and the depressions inside. This is the **Rayleigh-Taylor instability**.

3.7   TIME-DEPENDENT MAGNETIC FIELD

Here we treat slowly varying fields with $\partial/\partial t << \omega_C$, i.e., the variation of the respective field shall be small during one gyrocycle.

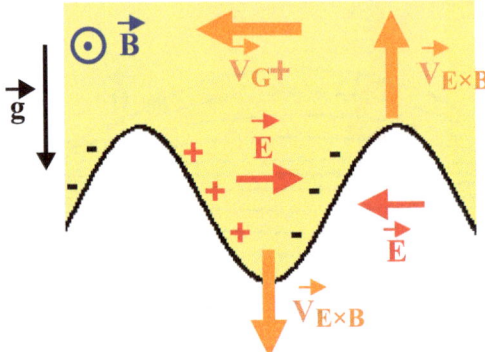

Figure 3.8: Rayleigh-Taylor instability of the plasma boundary due the gravitational drift.

Whereas a constant or time-dependent electric field directly alters the translational kinetic energy of a charged particle, a static magnetic field acts only indirectly on it, as described in Section 3.4.

According to the Maxwell equations, a time-dependent magnetic field results in a time-dependent electric field according to

$$\vec{\nabla} \times \vec{E} = -\frac{\partial \vec{B}}{\partial t} \tag{3.28}$$

or in integral form

$$\oint \vec{E} \, d\vec{l} = \int \left(\vec{\nabla} \times \vec{E} \right) d\vec{A} = -\frac{\partial}{\partial t} \int \vec{B} \, d\vec{A}, \tag{3.29}$$

where l is the line integration path around the area A.

Equation (3.28) is now applied to a gyrocircle in order to study the change of the energy of the gyration.

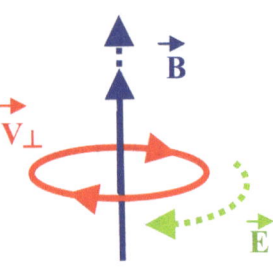

Figure 3.9: Induced electric field in the gyroplane for a slowly varying magnetic field.

As the electric field has only an azimuthal component as the vector of the perpendicular velocity, it accelerates or decelerates the particle continuously. The equation of motion is

$$m\frac{dv_\perp}{dt} = |q|E. \tag{3.30}$$

(Note that the sign of the charge does not enter as the azimuthal velocity changes also direction at inverse charge.) The change in gyrational energy is then calculated by

$$\frac{dW_\perp}{dt} = \frac{d}{dt}\left(\frac{m}{2}v_\perp^2\right) = mv_\perp\frac{dv_\perp}{dt} = |q|Ev_\perp \tag{3.31}$$

and using Eqs. (3.29), (3.5), and (3.16)

$$\frac{dW_\perp}{dt} = qv_\perp\frac{\pi r_L^2}{2\pi r_L}\frac{\partial B}{\partial t} = \frac{mv_\perp^2}{2B}\frac{\partial B}{\partial t} = \mu\frac{\partial B}{\partial t}. \tag{3.32}$$

Using Eq. (3.16) again,

$$\frac{d\mu}{dt} = \frac{1}{B}\frac{dW_\perp}{dt} - \frac{W_\perp}{B^2}\frac{\partial B}{\partial t} = 0. \tag{3.33}$$

Thus, the magnetic moment is not only conserved for a spatially varying static magnetic field as shown in Section 3.4, but also for a field which varies slowly in time. However, the gain (or loss) in transversal energy is not compensated by a loss (or gain) in translational energy, so that the total kinetic energy can be changed by a slowly varying magnetic field and the plasma can be heated (or cooled) in this way.

In an axially magnetized solenoidal plasma as shown in Fig. 3.8, Eq. (3.29) reads in cylindrical coordinates

$$2\pi rE_\theta = -\pi r^2\frac{\partial B_z}{\partial t}. \tag{3.34}$$

From this, an ExB drift results in radial direction with a velocity

$$v_r = -\frac{r}{2B_z}\frac{\partial B_z}{\partial t} \tag{3.35}$$

independent of the polarity, to that the whole plasma is moved inward. As $v_r \sim r$, this "compression" occurs homogeneously, with the plasma conserving its shape and its charged particle distribution. Denoting the plasma radius by r_p, the variation of the magnetic flux through the plasma during the compression is

$$\frac{d\Phi}{dt} = \frac{d}{dt}\left(\pi r_p^2 B_z\right) = 2\pi r_p B_z v_{rp} + \pi r_p^2\frac{\partial B_z}{\partial t}, \tag{3.36}$$

where v_{rp} denotes the inward velocity of the plasma circumference. According to Eq. (3.35),

$$\frac{d\Phi}{dt} = 0 \tag{3.37}$$

which means that the magnetic flux through the plasma is conserved during the compression phase, just as the magnetic flux through the gyro-orbit is conserved. In other words, the charged particles remain attached to their field lines during the compression. The particles are "**frozen**" to their field lines.

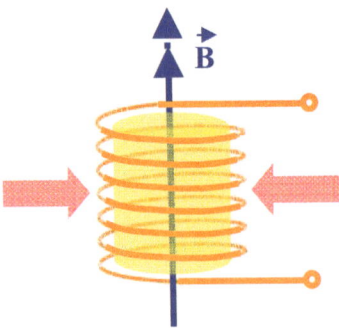

Figure 3.10: Radial compression of a solenoidal plasma in a slowly varying magnetic field.

3.8   TIME-DEPENDENT ELECTRIC FIELD

Let us assume the geometry of Fig. 3.3 with an electric field being perpendicular to the magnetic field, and varying slowly in strength. Then, by differentiating Eq. (3.6),

$$\frac{d^2 v_x}{dt^2} = -\omega_c^2 v_x \pm \frac{\omega_c}{B} \frac{\partial E_x}{\partial t}. \tag{3.38}$$

Separating v_x into a (fast) gyration oscillation and a (slow) constant drift velocity results in

$$\frac{d^2 v_{x,G}}{dt^2} = -\omega_c^2 v_{x,G} - \omega_c^2 v_{x,D} \pm \frac{\omega_c}{B} \frac{\partial E_x}{\partial t}, \tag{3.39}$$

from which the term on the left-hand side and the first term on the right-hand side cancel as they define the gyration. This results in the so-called "**polarization**" **drift** in the direction of the electric field,

$$\vec{v}_p = \frac{m}{qB^2} \frac{d\vec{E}_\perp}{dt}, \tag{3.40}$$

where $E_\perp$ denotes the electric field perpendicular to the magnetic field. The polarization drift separates again the charge carriers, and acts differently on ions and electrons (largely on the ions). Thus, it also induces a net current in the plasma (in analogy to the polarization current in dielectric materials.)

3.9   ADIABATIC INVARIANTS

Some of the above findings can be formulated in the concept of adiabatic invariance of classical mechanics. For a conservative (Hamiltonian) system, which is represented by the motion of charged particles in electric and magnetic fields when collisions are neglected, the integral of action

$$A = \oint \vec{p}\,d\vec{q} \tag{3.41}$$

is adiabatically constant on a path of a nearly periodic motion, where p denotes the momentum and q the location of the particle during the periodic motion. In our case, the nearly periodic motion can be, e.g., the gyration around the magnetic field. "Adiabatically" means that the fields change slowly, as described above.

For the gyration,

$$I_1 = \oint \vec{p}\,d\vec{q} = 2\pi m \upsilon_\perp r_L = 4\pi W_\perp \omega_C^{-1} = 4\pi \frac{m}{|q|}\mu \tag{3.42}$$

which is exactly the above conservation of the magnetic moment in static and slowly varying magnetic fields. This is called **the first adiabatic invariant** of motion in a magnetic field.

In a magnetic mirror configuration, as shown in Fig. 3.11, a charged particle is repelled from the high-field regions and thus performs an oscillatory axial motion along the direction of the magnetic field. The curvature drift is superimposed resulting in zig-zag azimuthal procession

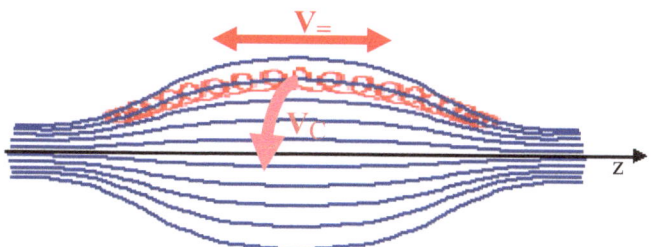

Figure 3.11: Axial pendulum motion of a charged particle in a magnetic mirror configuration.

of the axial trajectories. The so-called **second adiabatic invariant** is the loop integral of the parallel velocity during the pendulum motion (with the drift separated—see below)

$$I_2 = \oint m\vec{\upsilon}_=\,d\vec{s}, \tag{3.43}$$

where s denotes the path length of the pendulum motion. The general proof of the invariance is lengthy and will not be demonstrated here. We will, however, reformulate Eq. (3.43) into a relation for the magnetic field. According to Eqs. (3.21) and (3.23), energy conservation results in a total kinetic energy

$$W = W_= + \mu B. \tag{3.44}$$

Thus, the term μB acts like a mechanical potential during the pendulum motion. At the points of reversal, $W_=$ vanishes, so we obtain

$$W_=(s) + \mu B(s) = \mu B_m, \tag{3.45}$$

where B_m denotes the field at the points of reversal. Then from (3.43)

$$I_2 = \oint \sqrt{2mW_=(s)}\, ds = \oint \sqrt{2m\,(\mu B_m - \mu B(s))}\, ds. \tag{3.46}$$

In a static field, this expression remains apparently constant, at least in the axisymmetric configuration. When the magnetic field varies in time, however, the second adiabatic invariance does not hold. In accordance with the results of Section 3.6, the transversal energy is increased continuously, so that the magnetic moment will not be constant.

Finally, the azimuthal rotation, which is connected with the pendulum motion due to the curvature drift (see Fig. 3.12), leads to the third adiabatic invariant. As the resulting zig-zag trajectory of the gyrocenter proceeds on a shell along azimuthally identical field lines, the magnetic flux which is enclosed by a full azimuthal rotation is independent of the axial position z, and given by

$$I_3 = \oint m\vec{v}_C\, d\vec{s} = m \int \left(\vec{\nabla} \times \vec{v}_C\right) d\vec{A} \sim \int \vec{B}\, d\vec{A} = \Phi, \tag{3.47}$$

where s denotes here the path along a drift circle. The proof of the proportionality between I_3 and Φ is lengthy and is not given here.

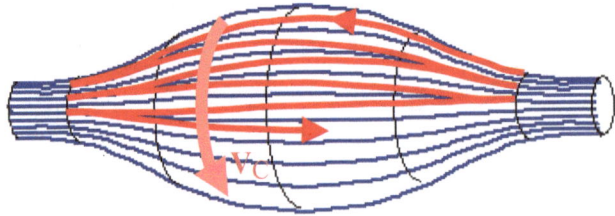

Figure 3.12: Zig-zag motion of the gyrocenter due to azimuthal curvature drift in a magnetic magnetic mirror configuration.

CHAPTER 4

Plasma as a Fluid

4.1 DISTRIBUTION FUNCTION AND MOMENTS

The single-particle descriptions of Chapters 2 and 3 illustrate the basic kinematical phenomena in a plasma, but are insuitable for a global description of the plasma dynamics due to the many particles involved. In fluid modeling, the species of the plasma are represented each by a distribution function f over the six-dimensional phase space, which changes with time. The distribution function is defined in such a way that the number of particles in a phase volume element $d^3 \times d^3 v$ is given by

$$d^6 N(t) = f(\vec{x}, \vec{v}, t) d^3 x d^3 v. \tag{4.1}$$

Integrating over the velocity space yields the particle (atomic) density

$$n(\vec{x}, t) = \int f(\vec{x}, \vec{v}, t) d^3 v. \tag{4.2}$$

A special distribution function of the velocity (or kinetic energy) is the Maxwellian distribution as presented in Chapter 1.

For a simplified two-dimensional representation of the phase space (Fig. 4.1), we consider the development of a volume element with time. During the time increment dt, the location

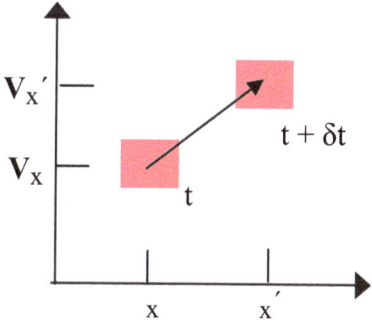

Figure 4.1: Displacement of a volume element in phase space during a time increment δt.

varies according to the velocity v and the force F,

$$x' = x + v_x \delta t \qquad v'_x = v_x + \frac{F}{m} \delta t. \tag{4.3}$$

Without particle loss or generation, the number of particles in the volume element remains constant

$$f(x', v'_x, t')dx' dv'_x = f(x, v_x, t)dx dv_x. \tag{4.4}$$

For the transformation, the Jacobi determinant results as

$$\frac{dx' dv'_x}{dx dv_x} = \begin{vmatrix} 1 & \delta t \\ 0 & 1 \end{vmatrix} = 1 \tag{4.5}$$

so that the distribution function remains constant in time (returning to 6D phase space)

$$\frac{d}{dt} f(\vec{x}, \vec{v}, t) = 0 \tag{4.6}$$

which is the **theorem of Liouville**. With the partial derivatives, this corresponds to

$$\frac{df}{dt} = \frac{\partial f}{\partial t} + \frac{\partial f}{\partial \vec{x}} \frac{\partial \vec{x}}{\partial t} + \frac{\partial f}{\partial \vec{v}} \frac{\partial \vec{v}}{\partial t} = 0 \tag{4.7}$$

from which the **Vlasov equation** results as

$$\frac{\partial f}{\partial t} + \vec{v} \frac{\partial f}{\partial \vec{x}} + \frac{\vec{F}}{m} \frac{\partial f}{\partial \vec{v}} = 0 \tag{4.8}$$

which describes the development of free, non-interacting particles. In general, collisions have to be taken into account, which will change the total time differential of f. This is incorporated in the following **Boltzmann equation** by a collision term, which includes all temporal changes of f due to collisions

$$\frac{\partial f}{\partial t} + \vec{v} \frac{\partial f}{\partial \vec{x}} + \frac{\vec{F}}{m} \frac{\partial f}{\partial \vec{v}} = \left(\frac{\partial f}{\partial t} \right)_{coll}. \tag{4.9}$$

The velocity dependence of f can be separated by the definition of moments, leaving a series of n-th moments which depend only on location

$$f^{(n)}(\vec{x}, t) = \int v^n f(\vec{x}, \vec{v}, t) d^3 v. \tag{4.10}$$

In particular, the zero-th moment is the density according to Eq. (4.2), and the first moment divided by the zero-th moment is the average velocity

$$\langle v \rangle = \frac{\int vf(\vec{x}, \vec{v}, t) d^3 v}{\int f(\vec{x}, \vec{v}, t) d^3 v} = \frac{1}{n} \int vf(\vec{x}, \vec{v}, t) d^3 v. \tag{4.11}$$

4.2   PARTICLE, MOMENTUM, AND ENERGY BALANCE

We now calculate the zero-th moment of the Vlasov equation according to

$$\frac{\partial}{\partial t}\int f d^3\upsilon + \int \vec{\upsilon}\frac{\partial f}{\partial \vec{x}}d^3\upsilon + \int \frac{\vec{F}}{m}\frac{\partial f}{\partial \vec{\upsilon}}d^3\upsilon = 0. \tag{4.12}$$

First, we look at the third term. For neutral particles, it vanishes if collisions and gravity are neglected. On charged particles, the Lorentz force acts, which we separate into the electric and magnetic field. For the electric field, which does not depend explicitly on the velocity in conservative systems, we obtain with the help of the Gauss integral relation

$$\int \vec{E}\vec{\nabla}_\upsilon f d^3\upsilon = \int \vec{\nabla}_\upsilon\cdot(\vec{E}f)d^3\upsilon = \oint f\vec{E}d^2\vec{\upsilon} = 0 \tag{4.13}$$

as the surface of the velocity space can be extended to infinity where the distribution function vanishes. For the magnetic force, the third term of Eq. (4.12) becomes by partial vector differentiation

$$\int (\vec{\upsilon}\times\vec{B})\vec{\nabla}_\upsilon f d^3\upsilon = \int \vec{\nabla}_\upsilon\cdot(f\vec{\upsilon}\times\vec{B})d^3\upsilon - \int f\vec{\nabla}_\upsilon\cdot(\vec{\upsilon}\times\vec{B})d^3\upsilon = 0 \tag{4.14}$$

as in the middle expression, the first term vanishes again by the application of the Gauss relation. In the second term, the differential operator is parallel to the velocity whereas the vector product is orthogonal to the velocity, so that the scalar product vanishes (which can easily be confirmed by expressing the div and the vector product in Cartesian coordinates). Thus, the third term of Eq. (4.12) vanishes. To the middle term, product differentiation is applied, so that

$$\frac{\partial}{\partial t}\int f d^3\upsilon + \frac{\partial}{\partial \vec{x}}\int \vec{\upsilon}f d^3\upsilon - \int f\frac{\partial}{\partial \vec{x}}\vec{\upsilon}d^3\upsilon = 0 \tag{4.15}$$

with the last term vanishing due to the derivative. Therefore, according to Eq. (4.11),

$$\frac{\partial n}{\partial t} + \frac{\partial}{\partial \vec{x}}(n\langle\vec{\upsilon}\rangle) = 0. \tag{4.16}$$

This is the well-known **equation of continuity**. It should be noted that for an isotropic velocity distribution, such as the Maxwell distribution, the average velocity vector vanishes, so that the density is conserved. Only a directed flux changes density. Integration Eq. (4.16) over a certain volume V yields for the number of particles N in that volume, again by the Gauss relation

$$\frac{\partial N}{\partial t} + \oint \vec{j}d\vec{A} = 0 \tag{4.17}$$

with the directed flux

$$\vec{j} = n\langle\vec{\upsilon}\rangle. \tag{4.18}$$

As indicated in Fig. 4.2, N increases for a directed influx of particles, rendering the scalar product in Eq. (4.17) negative. It should be noted that an external influx of particles is equivalent to the motion of the volume element with respect to an isotropic background, such as for drift motions. Then, the average velocity is given by the velocity of the center of gravity of V.

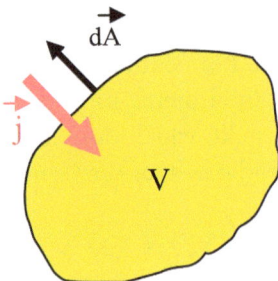

Figure 4.2: Directed influx (beam) of particles into the volume V.

Accordingly, we now separate the particle velocity within any volume element (see Fig. 4.3) and with respect to a coordinate system which is fixed in space into a directed velocity u (of the center of gravity) and the (isotropic) thermal velocity

$$\vec{v} = \vec{u} + \vec{v}_{th}. \tag{4.19}$$

By averaging, only u contributes to the first moment

$$\langle \vec{v} \rangle = \vec{u} \tag{4.20}$$

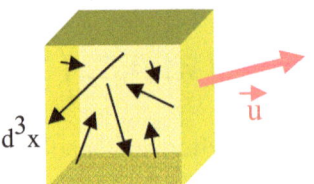

Figure 4.3: Velocity of the center of gravity of a volume element and thermal velocities of the particles.

Thus, the zero-th moment of the Vlasov equation defines the particle balance. The first moment is

$$\int m\vec{v}\frac{\partial f}{\partial t}d^3v + \int mv^2\frac{\partial f}{\partial \vec{x}}d^3v + \int \vec{v}\vec{F}\frac{\partial f}{\partial \vec{v}}d^3v = 0. \tag{4.21}$$

Here, we will not go through the lengthy details of the evaluation. Finally, with the microscopic definition of the pressure (see Eq. (1.3))

$$p = \frac{1}{3}nm\langle v_{th}^2\rangle \tag{4.22}$$

the **momentum balance** results as

$$mn\left[\frac{\partial \vec{u}}{\partial t} + (\vec{u}\cdot\vec{\nabla})\vec{u}\right] = n\vec{F} - \vec{\nabla}p = nq[\vec{E} + \vec{u}\times\vec{B}] - \vec{\nabla}p. \tag{4.23}$$

According to Eq. (4.19) and by averaging the thermal velocity over the distribution function, Eq. (4.23) represents the equation of motion of the center of gravity only. The left-hand side is the Newton equation of motion plus a drift term which arises from the motion of the center of gravity and is mostly small.

The second moment of the Vlasov equation reads

$$\int m\upsilon^2 \frac{\partial f}{\partial t}d^3\upsilon + \int m\upsilon^2\vec{\upsilon}\frac{\partial f}{\partial\vec{x}}d^3\upsilon + \int \upsilon^2\vec{F}\frac{\partial f}{\partial\vec{\upsilon}}d^3\upsilon = 0 \tag{4.24}$$

from which with the mean energy

$$\langle E \rangle = \frac{1}{2}m\langle \upsilon_{th}^2 \rangle \tag{4.25}$$

the **energy balance** results according to

$$\frac{\partial}{\partial t}n\langle E\rangle + \vec{\nabla}\cdot(n\langle E\rangle\vec{u}) + p\vec{\nabla}\cdot\vec{u} = \vec{\nabla}\cdot(\kappa\vec{\nabla}T) + n\vec{u}\vec{F}. \tag{4.26}$$

Here, the second term on the left-hand side is the convective transport of energy, and the third one describes compression (or expansion). On the right-hand side, the first term denotes the thermal conduction which is given by the thermal conductivity κ and the gradient of the temperature T. The last term denotes the energy gain or loss per unit volume due to external forces, i.e., the heating or cooling of the plasma.

4.3   DRIFTS IN FLUID DESCRIPTION

In the momentum balance, Eq. (4.23), the left-hand side, vanishes for a stationary system which constant velocity. For a velocity parallel to the magnetic field (or without a magnetic field) the result is

$$nq\vec{E} = \vec{\nabla}p \tag{4.27}$$

which reads for electrons with $q = -e$ and an electrostatic potential Φ

$$ne\vec{\nabla}\Phi = \vec{\nabla}nkT_e = kT_e\vec{\nabla}n \tag{4.28}$$

assuming that T_e is locally constant. The solution is

$$e\Phi = kT_e\log\frac{n}{n_e}, \tag{4.29}$$

where n_e is the integration constant defined in such a way that the electrostatic potential vanishes for $n = n_e$. Thus, Φ (and E) can be interpreted as to result from a fluctuation of the electron fluid over the background of the neutral plasma with electron density n_e. From Eq. (4.29), the **Boltzmann relation**

$$n = n_e \exp\left(\frac{e\Phi}{kT_e}\right) \tag{4.30}$$

results. As the ions are assumed at rest, the Poisson equation reads for small fluctuations

$$\nabla^2\Phi = -\frac{e}{\varepsilon_0}(n_e - n) = -\frac{en_e}{\varepsilon_0}\left(1 - \exp\left(\frac{e\Phi}{kT_e}\right)\right) \approx \frac{e^2 n_e}{\varepsilon_0}\frac{\Phi}{kT_e}. \tag{4.31}$$

The solution is

$$\Phi(x) = \Phi_0 \exp\left(-\frac{|x|}{\lambda_D}\right) \tag{4.32}$$

with the Debye length of Eq. (1.7). Thus, the fluid picture enables us to obtain a stringent derivation of the Debye length confirming the heuristic arguments of Section 1.3.

For a drift perpendicular to the magnetic field, we obtain from Eq. (4.23), again for a stationary plasma

$$nq\left[\vec{E} + \vec{u}_\perp \times \vec{B}\right] = \vec{\nabla}p. \tag{4.33}$$

Vector multiplication with B yields

$$nq[\vec{E} \times \vec{B} + \vec{B}(\vec{u}_\perp \cdot \vec{B}) - \vec{u}_\perp \vec{B}^2] = \vec{\nabla}p \times \vec{B}, \tag{4.34}$$

where the second term on the left-hand side vanishes as $u \perp B$. Therefore, the drift velocity becomes

$$\vec{u}_\perp = \frac{\vec{E} \times \vec{B}}{B^2} - \frac{\vec{\nabla}p \times \vec{B}}{nqB^2}. \tag{4.35}$$

In addition to Eq. (3.11), this includes the collective term according to the pressure gradient, which is called **diamagnetic drift**.

Figure 4.4 demonstrates the effect of the diamagnetic drift on a plasma in a cylindrical vessel with a longitudinal magnetic field and a radial pressure (or density) gradient of the charged particles. The diamagnetic drift causes a collective rotation of the charged particles. In the absence of an electric field, Eq. (4.35) yields the diamagnetic current density

$$\vec{j}_\perp = nq\vec{u}_\perp = \frac{\vec{B} \times \vec{\nabla}p}{B^2}. \tag{4.36}$$

Like the current in a conductive ring induced by a magnetic field which varies in time, the diamagnetic current tends to reduce the magnetic field in the plasma. (However, the present effect occurs in a static field.) Thus, the plasma tends to expel the field lines from its volume as displayed schematically in Fig. 4.5. For self-consistent description, the combination of Eq. (4.33) with the Maxwell equation

$$\vec{\nabla} \times \vec{B} = \mu_0 \vec{j}_\perp \tag{4.37}$$

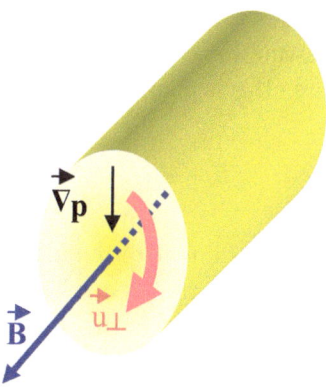

Figure 4.4: Diamagnetic rotation in a magnetized cylindrical plasma with a pressure gradient particles.

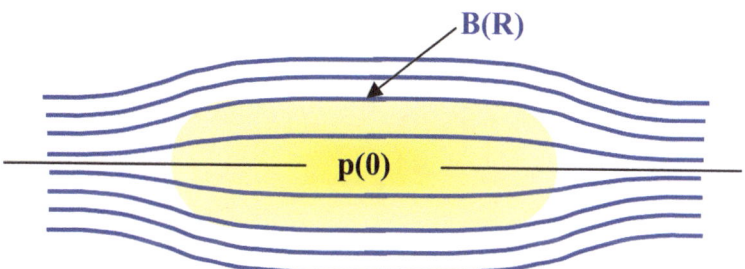

Figure 4.5: Expulsion of the magnetic field from a cylindrical plasma and the quantities determining the β value.

yields

$$\frac{1}{\mu_0}(\vec{\nabla} \times \vec{B}) \times \vec{B} = \frac{1}{\mu_0}\left((\vec{B} \cdot \vec{\nabla})\vec{B} - \frac{1}{2}\vec{\nabla}(B^2)\right) = \vec{\nabla}p \tag{4.38}$$

or

$$\vec{\nabla}\left(p + \frac{B^2}{2\mu_0}\right) = \frac{1}{\mu_0}(\vec{B} \cdot \vec{\nabla})\vec{B}. \tag{4.39}$$

In the often represented case of nearly parallel field lines (as, e.g., for the cylindrical plasma of Fig. 4.5), the term on the right-hand side is negligible, so that

$$p + \frac{B^2}{2\mu_0} = const. \tag{4.40}$$

Consequently, the quantity $B^2/2\mu_0$ is called the **magnetic pressure**. Applying Eq. (4.40) to the cross section of the cylindrical plasma and even outside, the magnetic field is reduced toward the center, which again is synonymous to the expulsion of the magnetic field lines from the plasma. In a plasma with high pressure (of charged particles), the reduction of the magnetic field is stronger than at low pressure. This is quantified by the so-called β **value** of the plasma, defined as (see Fig. 4.5)

$$\beta + \frac{p(0)}{\frac{B^2(R)}{2\mu_0}} \tag{4.41}$$

which compares the pressure at the center line with the magnetic field at the circumference. In "high-β" plasmas, the magnetic field within the plasma becomes very small.

The diamagnetic drift cannot be described in the single-particle picture (Chapter 3), as it is a collective effect. This is demonstrated in Fig. 4.6, where the density gradient across a volume element adds the gyrovelocities of many particles to a net drift velocity.

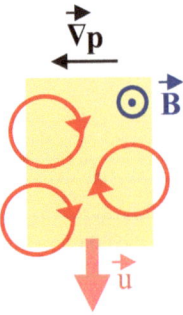

Figure 4.6: The diamagnetic drift velocity results from the gyration of many particles through a volume element, if a pressure gradient is present.

CHAPTER 5

Transport

5.1 DRIFT AND DIFFUSION

Our starting point is the momentum balance, Eq. (4.23), which is now expanded by a term which takes into account collisions with a collision frequency v_c (see Section 2.1). The simplest picture is that during the "collision" of a volume element with a background species the directed momentum relaxes completely, so that the momentum is lost. The collision frequency per unit volume is $n v_c$, so that in the presence of only an electric field

$$mn \left[\frac{\partial \vec{u}}{\partial t} + (\vec{u} \cdot \vec{\nabla})\vec{u} \right] = nq\vec{E} - \vec{\nabla}p - m\vec{u}v_c = 0. \tag{5.1}$$

In the stationary case the time derivative vanishes on the left-hand side. The second term can be neglected as the velocity becomes constant for a sufficiently high collision frequency. Then,

$$\vec{u} = \frac{1}{mnv_c}(nq\vec{E} - \vec{\nabla}p). \tag{5.2}$$

For isothermal variation of the pressure, the ideal gas equation is inserted. Furthermore, the multiplication with n converts into the particle flux

$$\vec{j} = \frac{1}{mv_c} \left(nq\vec{E} - kT\vec{\nabla}n \right). \tag{5.3}$$

The first term represents the flux of charged particles which is driven by the electric field. It is characterized by the **mobility**

$$\mu = \frac{|q|}{mv_c} \tag{5.4}$$

so that

$$\vec{j}_\pm = \pm \mu n \vec{E} \tag{5.5}$$

for charged particles of either polarity. For all kinds of particles, a diffusional flux

$$\vec{j}_d = -D\vec{\nabla}n \tag{5.6}$$

results with the **diffusion constant**

$$D = \frac{kT}{mv_c}. \tag{5.7}$$

Between the mobility and the diffusion constants, the **Einstein relation**

$$D = \frac{\mu k T}{|q|} \tag{5.8}$$

holds. Using the mean velocity of the Maxwell distribution and expressing the collision frequency by the mean free path length (Eqs. (1.3) and (2.3)), the diffusion constant kann also be written as

$$D = \frac{kT}{m} \frac{\lambda_c}{\langle v \rangle} = \sqrt{\frac{\pi}{8}} \lambda_c \sqrt{\frac{kT}{m}} = \frac{\pi}{8} \lambda_c \langle v \rangle = \frac{\pi}{8} v_c \lambda_c^2. \tag{5.9}$$

Following Eq. (5.6), the equation of continuity yields

$$\frac{\partial n}{\partial t} - \vec{\nabla} \cdot (D \vec{\nabla} n) = \frac{\partial n}{\partial t} - \vec{\nabla} D \cdot \vec{\nabla} n - D \nabla^2 n = 0. \tag{5.10}$$

Due to the high thermal conductivity of a plasma, it can generally be assumed that the temperature gradient is small. Then, D varies in space only through the density variation. From Eq. (5.9), $D \sim \lambda_C \sim n^{-1/3}$, this dependence is weak so that the second term in the middle can mostly be neglected, resulting in **Fick's law**

$$\frac{\partial n}{\partial t} = D \nabla^2 n. \tag{5.11}$$

5.2   TRANSPORT OF NEUTRALS

We first consider the diffusional transport regime where the findings of Section 5.1 are valid. For a plasma confined in a vessel with the linear dimension d, the diffusional regime applies provided the mean path length λ_c is small compared to d.

Density gradients, which drive the diffusion according to the above, can exist also for the parent gas of a plasma due to locally varying gas flows as, e.g., by the inlet and the outlet through a pump. Also the consumption by the plasma processes of ionization, dissociation, etc., may cause density gradients of the parent gas. Nevertheless, there are normally no strong gradients. On the other hand, for neutral species which are generated in the plasma, strong gradients will exist in particular when these species are reactive and adsorbed by the wall.

We consider a one-dimensional planar geometry with infinitely extended walls perpendicular to the direction of diffusion (see Fig. 5.1). With an ansatz of separation of variables according to

$$n(x,t) = v(t)w(x). \tag{5.12}$$

Equation (5.11) becomes

$$\frac{1}{v}\frac{dv}{dt} = \frac{D}{w}\frac{d^2w}{dx^2} = -\frac{1}{\tau}. \tag{5.13}$$

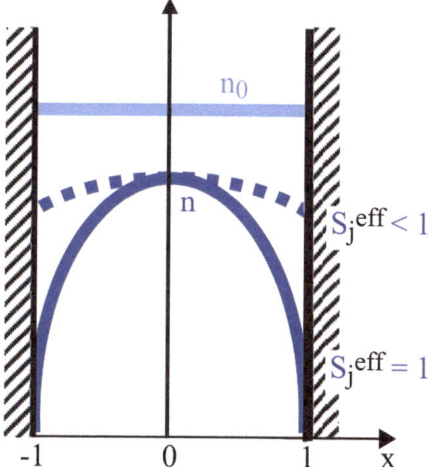

Figure 5.1: Schematic of a diffusional profile of a neutral species with unity and non-unity effective sticking. The parent gas density is schematically indicated by n_0.

As the left term depends only on t and the middle one only on x, both have to be constant with $-\tau^{-1}$ denoting the constant. The solution for υ is

$$\upsilon(t) = \upsilon_0 \exp\left(-\frac{t}{\tau}\right). \tag{5.14}$$

The spatial dependence becomes with

$$\Lambda = \sqrt{D\tau} \tag{5.15}$$

$$-\frac{w}{\Lambda^2} = \frac{d^2 w}{dx^2}. \tag{5.16}$$

The solution is

$$w = w_0 \cos\frac{x}{\Lambda} + w_1 \sin\frac{x}{\Lambda}. \tag{5.17}$$

The solution has to be symmetric so that the second term vanishes. The constants Λ and τ are obtained from the boundary conditions. For an ideally absorbing wall with the boundary conditions $n(x,t) = 0$ for $x = \pm l$ (see Fig. 5.1), the solution becomes

$$n(x,t) = n_{t=0} \cos\left(\frac{\pi}{2l}x\right) \exp\left(-\frac{\pi^2 Dt}{4l^2}\right). \tag{5.18}$$

This represents a decaying cosine half wave with a time constant

$$\tau = \frac{\Lambda^2}{D} = \frac{4l^2}{\pi^2}\frac{1}{D} \tag{5.19}$$

which is denoted as mean confinement time.

As the standard case, the respective neutral will diffuse in the background of the parent gas. As both species move thermally, the reduced mass (see Eq. (2.7)) has to be inserted into Eq. (5.9), so that the diffusion coefficient becomes

$$D = \sqrt{\frac{\pi}{8}} \lambda_c \sqrt{k T_0 \frac{m + m_0}{m \cdot m_0}}, \tag{5.20}$$

where T_0 denotes the gas temperature and m and m_0 the masses of the diffusing species and the parent gas atoms (molecules).

For different simple geometries, different functions are obtained for the diffusional density profiles which will not be considered here. Also, the "diffusional connection length" Λ changes. Table 5.1 lists these values for planar, cylindrical, and spherical geometries.

Table 5.1: Simplified geometries with definition of characteristic linear dimension and diffusional connection length

	Planar	Cylindrical	Spherical
d	1	$\rho/2$	$r/3$
Λ^2	$(2l/\pi)^2$	$\rho^2/5.8$	$(r/\pi)^2$

In this connection, it is also helpful to define the characteristic linear dimension according to

$$d = \frac{V}{A} \tag{5.21}$$

with V and A denoting the volume and the surface area of the container, respectively.

Neutrals are not necessarily completely absorbed by the walls. For example, a rare gas atom will hardly interact with the wall material at all. However, also reactive species may exhibit a wall attachment probability which is far below one, so that the major fraction is reemitted back into the plasma volume.

This is accounted for by the definition of the effective sticking coefficient

$$S_{eff} = \frac{j_{in} - j_{re}}{j_{in}}, \tag{5.22}$$

where j_{in} and j_{re} denote the incoming and reemitted neutral flux at the wall, respectively (see Fig. 5.2). If the effective sticking coefficient is less than 1, the neutral density at the wall is nonzero, as indicated in Fig. 5.1. More complicated calculations then yield a corrected connection length according to

$$\Lambda_{\textit{eff}}^2 \approx \Lambda^2 + \frac{2}{3}\lambda_c \frac{2 - S_{\textit{eff}}}{S_{\textit{eff}}} d. \tag{5.23}$$

In the limit $S_{\textit{eff}} \to 0$, the connection length becomes infinite which means that the profile remains flat. Then, the mean confinement time also becomes infinite as no loss occurs.

Figure 5.2: Incoming and reemitted neutral fluxes at a wall.

At low pressures below about 0.1 Pa, the mean free path length exceeds the characteristic linear dimension of typical laboratory devices. Then, any neutral in the plasma will just fly to the wall without any interaction with the remaining gas. This is called the free fall regime with $\lambda_c > d$. Then, the mean confinement time is inversely proportional to the mean thermal velocity

$$\tau = \frac{\Lambda}{\langle v \rangle} \tag{5.24}$$

with a free fall connection length

$$\Lambda = 2d. \tag{5.25}$$

(The factor of two accounts for the average path length, as the isotropic velocity distribution hits the wall.) Again, the connection length can be corrected for non-unity sticking by means of

$$\Lambda_{\textit{eff}} = \Lambda \frac{2 - S_{\textit{eff}}}{S_{\textit{eff}}}. \tag{5.26}$$

5.3 AMBIPOLAR DIFFUSION

The following considerations are related to partially ionized plasmas, where there is a large probability that the charged particles collide with the parent background gas.

In contrast to the diffusion of neutrals, the diffusion of the charged species in a plasma, ions and electrons, is expected to be strongly asymmetric, as often the electron temperature is much higher, and the masses are widely different. From this, the electron mobility is much higher, so that the electron density profile would decay much faster due to diffusion. This, however, would violate quasineutrality, and electric fields will result which counteract the charge separation. To maintain quasineutrality, the diffusive fluxes of positive ions and electrons have to be equal, so that from Eq. (5.3)

$$\mu_i n_i \vec{E} - D_i \vec{\nabla} n_i = -\mu_e n_e \vec{E} - D_e \vec{\nabla} n_e. \tag{5.27}$$

Due to neutrality, the local densities of ions and electrons, as well as their gradient, are equal as we consider distances which are much larger than the Debye length. Thus, the so-called ambipolar field becomes

$$\vec{E} = \frac{D_i - D_e}{\mu_i + \mu_e} \frac{\vec{\nabla} n_e}{n_e} \tag{5.28}$$

and the diffusive flux of charges

$$\vec{j} = \mu_i n_e \frac{D_i - D_e}{\mu_i + \mu_e} \frac{\vec{\nabla} n_e}{n_e} - D_i \vec{\nabla} n_e = -\frac{\mu_i D_e + \mu_e D_i}{\mu_i + \mu_e} \vec{\nabla} n_e. \tag{5.29}$$

This defines the ambipolar diffusion coefficient

$$D_a = \frac{\mu_i D_e + \mu_e D_i}{\mu_i + \mu_e}. \tag{5.30}$$

As $\mu_e \gg \mu_i$ and with the Einstein relation (Eq. (5.8)),

$$D_a = D_i + \frac{\mu_i}{\mu_e} D_e = D_i \left(1 + \frac{k T_e}{k T_i} \right). \tag{5.31}$$

In non-thermal plasmas, when the ion temperature is small compared to the electron temperature, the field-driven term dominates in Eq. (5.3) for the diffusion of ions so that

$$D_a = D_i \frac{k T_e}{k T_i} = \frac{\mu_i}{e} k T_e = \frac{k T_e}{m_i \nu_{ci}}. \tag{5.32}$$

This shows that the charged particle diffusion is given by the inertia of the ion collision dynamics, but driven by the electron temperature. On the other hand, due to the high mobility of the electrons, the gradient of the electron density is small and the transport of the electrons is governed by their thermal velocity. Thereby, electrons reach the wall by their thermal motion, and pull the inert ions behind them.

This picture will be further elucidated when treating below the particle fluxes at the plasma boundary.

5.4 DIFFUSION IN A MAGNETIC FIELD

So far, the magnetic field has been neglected. However, due to the vector product in the momentum balance, the above treatment is valid as well for the transport along the magnetic field lines.

For the diffusional transport perpendicular to the magnetic field, we here neglect drifts as described in Chapters 3 and 4. Then, a transport perpendicular to the field lines can only be accomplished by collisions which causes jumps of the gyrating particles to adjacent field lines.

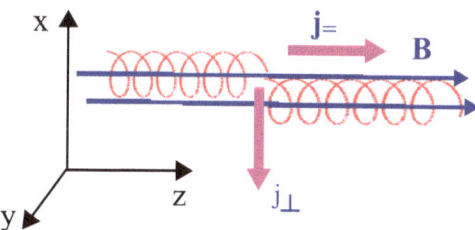

Figure 5.3: Transport parallel and perpendicular to the magnetic field lines. The cross-field transport is governed by collisions.

The cross-field diffusion is then given by a random walk of the guiding center of the gyration. The momentum balance Eq. (4.23) then reads with the collision term added

$$mn\frac{\partial \vec{u}_\perp}{\partial t} = nq[\vec{E} + \vec{u}_\perp \times \vec{B}] - kT\vec{\nabla}n - mn\nu_c\vec{u}. \tag{5.33}$$

The left-hand side describes the gyration. Being only interested in the drifts, we neglect it again. Putting the magnetic field and the direction of cross-field transport in the z and x directions, respectively, Eq. (5.33) becomes

$$mn\nu_c u_x = \pm en E_x \pm en\nu_y B - kT\frac{\partial n}{\partial x}$$

$$mn\nu_c u_y = \pm en E_y \mp en\nu_x B - kT\frac{\partial n}{\partial y}. \tag{5.34}$$

Introducing mobility, Einstein relation and cyclotron frequency, we obtain

$$u_x = \pm\mu E_x \pm \frac{\omega_C}{\nu_c}\nu_y - \frac{D}{n}\frac{\partial n}{\partial x} \qquad u_y = \pm\mu E_y \mp \frac{\omega_C}{\nu_c}\nu_x - \frac{D}{n}\frac{\partial n}{\partial y}. \tag{5.35}$$

Inserting into each other yields with the mean collision time $\tau_c = 1/\nu_c$

$$u_x(1 + \omega_C^2\tau_c^2) = \pm\mu E_x + \omega_C^2\tau_c^2\frac{E_y}{B} \mp \omega_C^2\tau_c^2\frac{kT}{eB}\frac{1}{n}\frac{\partial n}{\partial y} - \frac{D}{n}\frac{\partial n}{\partial x}. \tag{5.36}$$

The first and last term denote the well-known drift and diffusion. The second term represents the ExB drift, and the third one corresponds to the diamagnetic drift. Formally, the diffusional drift is thus modified by the bracket on the left-hand side so that the cross-field mobility and diffusion coefficient can be written as

$$\mu_\perp = \frac{\mu_=}{1 + \omega_C^2\tau_c^2} \tag{5.37}$$

and

$$D_\perp = \frac{D_=}{1 + \omega_C^2\tau_c^2}. \tag{5.38}$$

Mobility and diffusion are retarded perpendicular to the field at decreasing collision frequency—confirming that the cross-field diffusion is mediated by collisions and increasing cyclotron frequency—which increases at increasing magnetic field. At sufficiently high magnetic field and inserting the ambipolar diffusion coefficient, Eq. (5.31), the diffusion coefficient becomes

$$D_\perp = \frac{kT_e}{m_i \nu_c} \frac{\nu_c^2}{\omega_C^2}.$$ (5.39)

Replacing the cyclotron frequency by the Larmor radius and the velocity perpendicular to the magnetic field, which is proportional to the square root of the temperature, yields

$$D_\perp \sim \nu_c r_L^2$$ (5.40)

which, in comparison to Eq. (5.9), can be understood as a random walk with the Larmor radius as the mean free path length. This is consistent with the fact that the Larmor radius is small compared to the mean path length in most magnetized plasmas.

5.5   PLASMA RESISTIVITY

Due to their high mobility, the electrical current through the plasma is carried by the electrons. Converting the particle flux of Eq. (5.5) into an electrical current density j_{el} and making use of the usual relation

$$\vec{j}_{el} = \frac{1}{\eta} \vec{E}$$ (5.41)

the DC resistivity of the plasma results as

$$\eta_{DC} = \frac{m_e c_{ce}}{n_e e^2} = \frac{\nu_{ce}}{\varepsilon_0} \frac{1}{\omega_{pe}^2},$$ (5.42)

where ν_{ce} denotes the mean collision frequency of the electrons.

As the current transport is one-dimensional, an average collision frequency cannot just be obtained by averaging three-dimensionally over the Maxwell distribution, Eq. (1.2). Instead, the one-dimensional average in the direction of the drift velocity u can be defined by

$$\nu_{ce}\vec{u} = \langle \nu_{ce}\vec{\upsilon} \rangle,$$ (5.43)

where υ is split into the drift and the thermal motion according to Eq. (4.19). Then, the velocity distribution over υ is "shifted" by the drift velocity according to

$$f_e(\upsilon) = \left(\frac{m_e}{2\pi kT_e} \right)^{3/2} \exp\left(-m_e \frac{|\vec{\upsilon} - \vec{u}|^2}{2kT_e} \right).$$ (5.44)

As $u \ll \upsilon$, the exponential becomes

$$\exp\left(-\frac{m_e(\upsilon^2 + u^2)}{2kT_e} \right) \exp\left(\frac{m_e \vec{u} \cdot \vec{\upsilon}}{kT_e} \right) = \left(1 + \frac{m_e \vec{u} \cdot \vec{\upsilon}}{kT_e} \right) \exp\left(-\frac{m_e \upsilon^2}{2kT_e} \right).$$ (5.45)

Thus, f becomes a modified unshifted Maxwellian (drift in z direction)

$$f_e(\upsilon) = \left(1 + \frac{m_e u_z \upsilon_z}{kT_e}\right) f_{eM}(\upsilon). \tag{5.46}$$

Performing the averaging on the right side of Eq. (5.43) with this distribution function, yields

$$\langle \upsilon_{ce} \upsilon_z \rangle = \int f_e(\upsilon) \upsilon_{ce} \upsilon_z d^3\upsilon = \frac{m_e u_z}{kT_e} \int f_{eM}(\upsilon) \upsilon_{ce} \upsilon_z^2 d^3\upsilon \tag{5.47}$$

as the first term in Eq. (5.46) vanishes upon integration due to the antisymmetry of υ_z. As $\upsilon^2 = \upsilon_x^2 + \upsilon_y^2 + \upsilon_z^2$, and the velocity is isotropic, we obtain according to Eq. (5.43),

$$\nu_{ce} = \frac{m_e}{3kT_e} \int f_{eM}(\upsilon) \upsilon_{ce} \upsilon^2 d^3\upsilon. \tag{5.48}$$

In a cold plasma with a low degree of ionization, electrons collide mainly with the neutral gas atoms, with elastic collisions dominating (Section 2.8, see also Fig. 2.13), so that no analytic expression is available.

In a fully singly ionized plasma, only Coulomb collisions occur between the electrons and the ions. Taking the 90° scattering as characteristic for the scattering cross section, Eq. (2.37) yields with $m_1 \ll m_2$

$$\sigma_{ei} = \frac{32}{\pi} \frac{e^4 \log \Lambda}{(4\pi\varepsilon_0)^2 m_e^2 \upsilon_e^4} \tag{5.49}$$

from which

$$\nu_{ei} = n_e \upsilon_e \sigma_{ei} = \frac{32}{\pi} \frac{n_e e^4 \log \Lambda}{(4\pi\varepsilon_0)^2 m_e^2 \upsilon_e^3}. \tag{5.50}$$

Inserting this into Eq. (5.48), the velocity integration yields

$$\int f_{eM}(\upsilon) \frac{d^3\upsilon}{\upsilon} = 4\pi \int f_{eM}(\upsilon) \upsilon d^3\upsilon = \sqrt{\frac{2m_e}{\pi kT_e}}. \tag{5.51}$$

Putting together Eqs. (5.40), (5.46), (5.48), and (5.49), the resistivity in the fully ionized plasma results as

$$\eta_{ei} = \frac{32}{3\pi^{3/2}} \frac{(2m_e)^{1/2}}{(kT_e)^{3/2}} \frac{e^2}{(4\pi\varepsilon_0)^2} \log \Lambda \tag{5.52}$$

which is independent of electron density, as the electron density is equal to the ion density which causes the scattering.

5.6 ELECTRICAL PLASMA HEATING

Plasma heating is mostly accomplished by accelerating the high-mobility electrons in an electric field, which will then be thermalized due to collisions. If a homogeneous constant electric field is applied across a plasma with volume V, the power per unit volume is given by (see Eqs. (5.41) and (5.42))

$$P_{DC} = \vec{j}_{el}\vec{E} = \frac{\varepsilon_0 \omega_{pe}^2}{v_{ce}}|\vec{E}|^2. \tag{5.53}$$

In an alternating electric field with amplitude E_0 and angular frequency ω, the momentum balance reads

$$n_e m_e \ddot{x} + n_e m_e \dot{x} v_{ce} = -n_e e E_0 e^{i\omega t}. \tag{5.54}$$

Setting $x = x_0 e^{i\omega t}$, Eq. (5.54) results in

$$j_{el} = -n_e e \dot{x} = -\varepsilon_0 \omega_{pe}^2 \frac{i - v_{ce}/\omega}{\omega(1 + v_{ce}^2/\omega^2)} E_0 e^{i\omega t}. \tag{5.55}$$

From this, the time-averaged power per unit volume is obtained according to

$$P_{AC} = \frac{1}{T}\int_0^T j_{el}(t) \cdot E(t)\,dt = \frac{1}{2}Re(j_{el} \cdot E^*) \tag{5.56}$$

resulting in

$$P_{AC} = \frac{\varepsilon_0 \omega_{pe}^2}{v_{ce}} \frac{v_{ce}^2}{v_{ce}^2 + \omega^2} \frac{E_0^2}{2}. \tag{5.57}$$

As the last term represents the square of the effective AC field, the AC resistivity results in comparison to the DC one (Eq. (5.42))

$$\eta_{AC} = \eta_{DC} \frac{v_{ce}^2 + \omega^2}{v_{ce}^2}. \tag{5.58}$$

Figure 5.4 evaluates Eq. (5.57). At low electron collision frequency corresponding to low gas density, the absorbed power tends to zero. Qualitatively, the electrons just follow the AC field without any net energy gain, unless collisions push them out of phase. At given angular frequency of the field, the absorption is maximum when the collision frequency matches the angular frequency. At given electric field, the maximum absorbed power decreases inversely proportional to the angular frequency. For sufficiently large collision frequency, the AC power absorption is more efficient than the DC one. In accordance with Eq. (5.58), it approaches the DC value in the limit of high collision frequency.

It should be noted that the above dependence on the angular frequency holds for a fixed AC amplitude. For a special arrangements of AC coupling, different dependencies may result (see, e.g., Chapter 12).

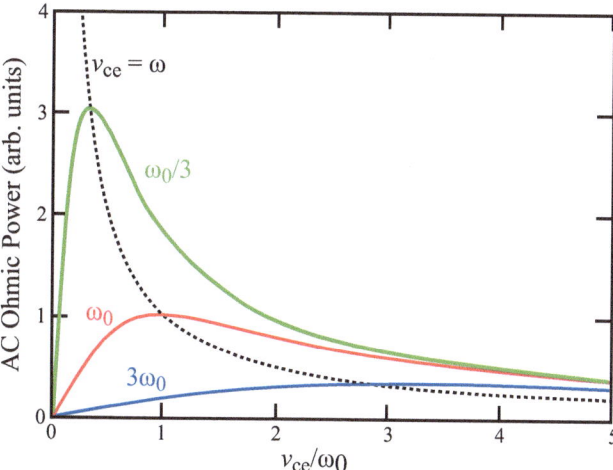

Figure 5.4: Ohmic heating power in an alternating electric field as function of the electron collision frequency ν_{ce}, for three different angular frequencies which differ by a factor of 3. The dotted line indicates maximum absorption when the collision frequency equals the angular frequency.

Furthermore, the propagation of electromagnetic waves may strongly influence the plasma heating. This will be covered in Chapter 9.

CHAPTER 6

Plasma Boundary

6.1 ELECTROSTATIC SHEATH

We consider a plasma in front of a wall, which, e.g., is part of the vessel containing the plasma, or may serve as a substrate for technical plasma processes of surface treatment. As the electrons generally exhibit much larger velocities than the ions, they will preferentially leave the plasma and charge the plasma slightly positive. Thereby, an electrostatic potential develops between the plasma and the wall, which repels the electrons so that only electrons of the high-energy tail of the distribution function still reach the wall. Simultaneously, the ions are accelerated and a new stationary situation is established with a few high-energy electrons and many accelerated ions reaching the wall. Still, the ions are slow compared to the electrons, so that the fluxes balance and the quasi-neutrality in the plasma is not disturbed. However, the ions and electrons in the sheath do no longer obey the quasi-neutrality. This situation is schematically depicted in Fig. 6.1.

Denoting the velocity of the ions in the sheath entering the sheath by v_0, their energy conservation reads (always assuming a collisionless sheath)

$$\frac{m_i}{2}(v_i(x))^2 + e\Phi(x) = \frac{m_i}{2}v_{i0}^2 \tag{6.1}$$

with the electrostatic potential Φ which is normalized to 0 at the sheath boundary, and the velocity v_{i0} when entering the sheath. The continuity equation requires

$$n_i(x)v_i(x) = n_{i0}v_{i0}, \tag{6.2}$$

where n_{i0} denotes the ion density at the sheath boundary. Inserting Eq. (6.1) into Eq. (6.2) yields

$$n_i(x) = n_{i0}\left(1 - \frac{2e\Phi(x)}{m_i v_{i0}^2}\right)^{-1/2}. \tag{6.3}$$

For the electrons, Eq. (4.30) is applied with

$$n_e(x) = n_{e0}\exp\left(\frac{e\Phi(x)}{kT_e}\right). \tag{6.4}$$

By definition, the plasma at the sheath entrance is still quasineutral (actually, this defines the sheath boundary), so that $n_{i0} = n_{e0}$. The Poisson equation then yields

$$\frac{d^2\Phi(x)}{dx^2} = \frac{en_0}{\varepsilon_0}\left[\exp\left(\frac{e\Phi(x)}{kT_e}\right) - \left(1 - \frac{2e\Phi(x)}{m_i v_{i0}^2}\right)^{-1/2}\right]. \tag{6.5}$$

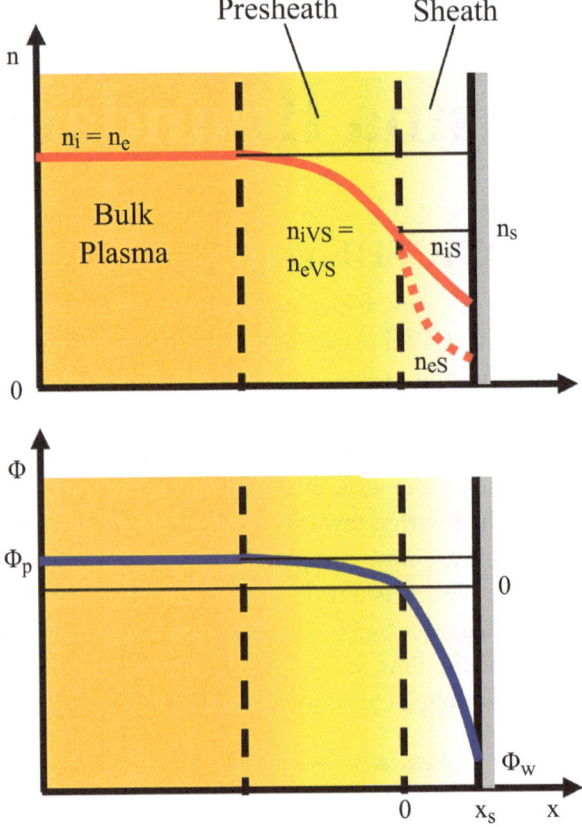

Figure 6.1: Schematic of the electron and ion densities and the electrostatic potential in front of a wall, defining the sheath and the presheath.

Changing variable to z, multiplying with $d\Phi/dz$ and integration yields on the left-hand side by product integration

$$\int_0^x \frac{d\Phi(z)}{dz}\frac{d^2\Phi(z)}{dz^2}dz = \frac{1}{2}\left[\left(\frac{d\Phi(z)}{dz}\right)^2\right]_0^x = \frac{1}{2}\left(\frac{d\Phi(x)}{dx}\right)^2 \tag{6.6}$$

as the potential has to be continuously differentiable at $x = 0$. On the right-hand side, z is substituted by $\Phi(z)$, which yields with $\Phi(0) = 0$

$$\frac{en_0}{\varepsilon_0}\int_0^x \frac{d\Phi(z)}{dz}[\ldots]\,dz = \frac{en_0}{\varepsilon_0}\int_0^{\Phi(x)}[\ldots]\,d\Phi \tag{6.7}$$

which can now be integrated with the result

$$\frac{1}{2}\left(\frac{d\Phi(x)}{dx}\right)^2 = \frac{en_0}{\varepsilon_0}\left[kT_e\exp\left(\frac{e\Phi(x)}{kT_e}\right) - kT_e + m_i v_{i0}^2\left(1 - \frac{2e\Phi(x)}{m_i v_{i0}^2}\right)^{1/2} - m_i v_{i0}^2\right]. \quad (6.8)$$

The right-hand side of this equation must not be negative. Sufficiently close to the sheath boundary at $x = 0$, $\Phi(x)$ is small. Correspondingly, the exponential and square root functions are expanded to second order, which yields for the expression in brackets

$$kT_e\left(\frac{e\Phi}{kT_e} + \frac{1}{2}\left(\frac{e\Phi}{kT_e}\right)^2\right) - m_i v_{i0}^2\left(\frac{e\Phi}{m_i v_{i0}^2} + \frac{1}{2}\left(\frac{e\Phi}{m_i v_{i0}^2}\right)^2\right) \geq 0 \quad (6.9)$$

or

$$\frac{(e\Phi)^2}{kT_e} \geq \frac{(e\Phi)^2}{m_i v_{i0}^2} \quad (6.10)$$

and finally

$$v_{i0} \geq v_B = \sqrt{\frac{kt_e}{m_i}}. \quad (6.11)$$

Thus, the conservation laws and electrostatics can only be fulfilled if the ions enter the sheath with a velocity which is equal or larger than the "Bohm" or "ion sound" velocity v_B. Equation (6.11) is also called the Bohm criterion. In a low-pressure pressure plasma with $kT_i \ll kT_e$, this velocity is much larger than the ion thermal velocity. More rigorous calculations show even that Eq. (6.11) holds even marginally, i.e.,

$$v_{i0} = v_B. \quad (6.12)$$

The exact solution of the sheath potential equation (6.5) can only be obtained numerically. From this, the thickness of the sheath is obtained as

$$x_s = (2\dots5)\cdot\lambda_D. \quad (6.13)$$

6.2   PRESHEATH

The existence of the Bohm velocity requires that the ions are accelerated on their way from the bulk plasma to the sheath. From this, a presheath (see Fig. 6.1) must exist in which the potential of the plasma drops to provide the ion acceleration. Above, we have defined the sheath boundary in such a way that quasineutrality extends to there, i.e., holds across the presheath. Then, from $n_i(x) = n_e(x)$

$$\frac{1}{n_i(x)}\frac{dn_i(x)}{dx} = \frac{1}{n_e(x)}\frac{dn_e(x)}{dx}. \quad (6.14)$$

Introducing the ion flux $j_i = en_i v_i$ and the Boltzmann relation for the electron density, we obtain

$$\frac{e v_i(x)}{j_i(x)} \frac{d}{dx} \frac{j_i(x)}{e v_i(x)} = \frac{1}{j_i(x)} \frac{dj_i(x)}{dx} - \frac{1}{v_i(x)} \frac{dv_i(x)}{dx} = \frac{e}{kT} \frac{d\Phi(x)}{dx}. \tag{6.15}$$

Within the presheath, v_i is smaller than the Bohm velocity. Thereby

$$\frac{1}{v_B} \frac{dv_i(x)}{dx} + \frac{e}{kT} \frac{d\Phi(x)}{dx} \leq \frac{1}{j_i(x)} \frac{dj_i(x)}{dx}. \tag{6.16}$$

The second term is negative and provides the acceleration of the ions,

$$-\frac{e}{kT} \frac{d\Phi(x)}{dx} = \frac{1}{kT} \frac{m_i}{2} \frac{d}{dx} v_i(x)^2 = \frac{1}{v_B} \frac{v_i(x)}{v_B} \frac{dv_i(x)}{dx}. \tag{6.17}$$

As again $v_i < v_B$, this is smaller than the first term in Eq. (6.16), so that the left-hand side is positive. Therefore, a contradiction arises as the ion flux is normally conserved, so that the right-hand side vanishes.

This can be resolved by:

(i) a collisional presheath which reduces the effective velocity gain so that the first term becomes smaller;

(ii) a geometrical presheath. For certain geometries, the ion flux can be compressed during the transport to the wall resulting in a positive right-hand side. This is the case for a convex cylinder or a sphere which are immersed into the plasma; and

(iii) an ionizing presheath which again increases the ion flux and turns the right-hand side positive.

6.3   POTENTIAL, FLUX, ION ENERGY

The plasma potential which is defined with respect to the sheath boundary (see Fig. 6.1) is easily calculated from the acceleration to the Bohm velocity

$$\frac{m_i}{2} v_B^2 = e\Phi_p \tag{6.18}$$

to

$$\Phi_p = \frac{kT_e}{2e}. \tag{6.19}$$

According to the Boltzmann relation, the electron (and ion) density at the sheath boudary, $x = 0$, is then reduced with respect to the bulk density and results as

$$n_{es} = n_e \exp(-1/2) \approx 0.6 n_e. \tag{6.20}$$

Then, the ion flux at the sheath boundary (and also to the wall) is given by

$$j_i = j_B = n_e v_B \exp(-1/2). \tag{6.21}$$

(It should be noted that the factor $e^{-1/2}$ is often not taken into account in literature). In low-pressure plasmas, the diffusive or a free-fall transport of the ions is fast so that the flux to the wall is indeed limited by the Bohm flux according to Eq. (6.21).

For the electron flux into the sheath, the isotropic thermal flux is assumed which is transmitted into the half-space

$$j_{es} = \frac{1}{4} n_e \langle v_e \rangle \exp(-1/2). \tag{6.22}$$

For a floating wall through which no electrical current is transported, the electron and ion flux must be equal. The electron flux at the wall is derived from Eq. (6.21) using the Boltzmann relation, Eq. (4.30), with the wall potential Φ_w. Thereby

$$\sqrt{\frac{kT_e}{m_i}} = \frac{1}{4} \sqrt{\frac{8kT_e}{\pi m_e}} \exp\left(\frac{e\Phi_w}{kT_e}\right) \tag{6.23}$$

and

$$\Phi_w = -\frac{kT_e}{2e} \log \frac{m_i}{2\pi m_e}. \tag{6.24}$$

Combining Eqs. (6.18) and (6.24), the floating potential between the bulk plasma and the wall results as

$$\Phi_{fl} = \Phi_p - \Phi_w = \frac{kT_e}{2e} \left(1 + \log \frac{m_i}{2\pi m_e}\right). \tag{6.25}$$

As schematically indicated in Fig. 6.2, the floating potential establishes in the plasma when no net current is drawn to the wall. The floating potential could be measured with a currentless probe immersed into the plasma, with respect to the chamber wall.

Correspondingly, the energy of the ions which impinge on the wall is in the floating situation

$$E_{i,fl} = e\Phi_{fl} = \frac{kT_e}{2} \left(1 + \log \frac{m_i}{2\pi m_e}\right). \tag{6.26}$$

Thus, the wall is bombarded by nonthermal ions with an energy in the order of 10 eV even if no external voltage is applied to the plasma.

6.4 NEGATIVELY BIASED EELECTRODE

For surface treatment, often a higher ion energy is required as resulting from the floating potential. This can be achieved on limited areas of the wall by applying negative DC bias (denoted by V_0—see Fig. 6.2) to an electrode immersed into the plasma. The electrode has to be small, as otherwise the plasma potential will be influenced by the bias. The negative bias will repel the

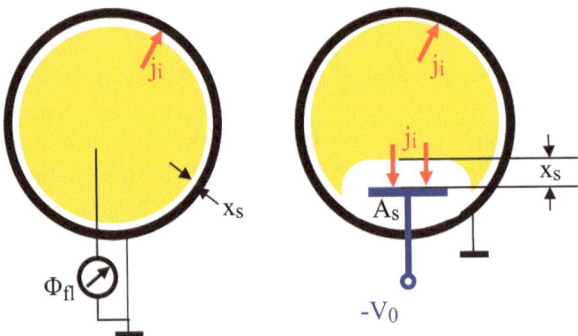

Figure 6.2: Schematic display of a floating plasma (left) and a plasma with an electrode with applied negative bias voltage.

electrons and accelerate the ions further. As we will see below, a new plasma boundary is formed above the electrode, for which still the above considerations hold. Therefore, the stationary ion flux is still given by the Bohm flux of Eq. (6.21) and cannot be influenced by the bias (if possible smaller geometrical effects are neglected). Rather than changing the ion flux, the sheath thickness is adjusted to accommodate the applied bias.

The dynamics of the boundary after a sudden application of the bias with $eV_0 >> kT_e$ is schematically indicated in Fig. 6.3. The electrons will be instantaneously repelled from the electrode region (actually, on a time scale of the inverse plasma frequency—see Section 1.4). The ions of density n_e (we here neglect the presheath) are inert and remain forming the so-called ion matrix sheath.

At sufficiently large bias, the floating potential can be neglected and the plasma is practically grounded. Then, the constant ion density across the matrix sheath results in the solution of the Poisson equation

$$\Phi(x) = -\frac{en_e}{\varepsilon_0}\frac{x^2}{2}, \tag{6.27}$$

where the origin of x is again the plasma boundary. The thickness of the matrix sheath results with $\Phi(x_s) = -V_0$ as

$$x_{sm} = \left(\frac{2\varepsilon_0 V_0}{en_e}\right)^{1/2} = \lambda_D \left(\frac{2eV_0}{kT_e}\right)^{1/2}. \tag{6.28}$$

Under the action of the applied bias, the ions of the matrix sheath are extracted from it. As an estimate for the characteristic time of matrix extraction, we take an ion from the sheath edge and assume that the potential, Eq. (6.27), will not change. (Actually, the dependence on x changes from quadratic to linear toward the empty matrix sheath.) Then, the equation of motion is

$$m_i \frac{d^2 x}{dt^2} = -e\frac{d}{dx}\Phi(x) = \frac{e^2 n_e}{\varepsilon_0}x. \tag{6.29}$$

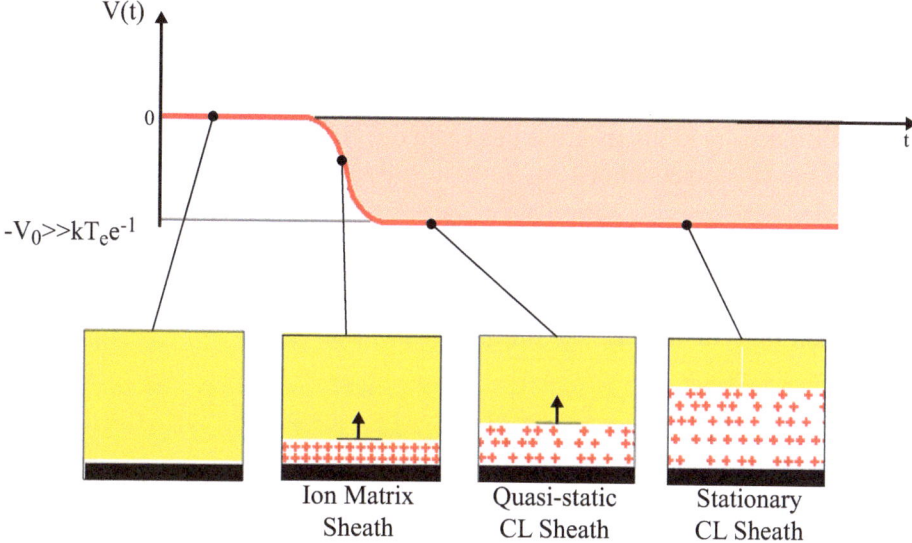

Figure 6.3: Formation of the ion matrix sheath and the Child-Langmuir sheath when negative bias is applied to an electrode immersed into the plasma.

The general solution is $x = A \cosh \omega_{pi} t + B \sinh \omega_{pi} t$ with

$$\omega_{pi} = \sqrt{\frac{e^2 n_e}{\varepsilon_0 m_i}} \qquad (6.30)$$

denoting the ion plasma frequency. For an ion starting at the plasma boundary, the initial condition $x = 0$ at $t = 0$ requires $A = 0$. For a nontrivial solution, an initial velocity is required which is given by the Bohm velocity, Eq. (6.11), so that $v = v_B$ at $t = 0$ yields $v_B = B \omega_{pi}$. Then, with Eq. (1.7), the solution becomes

$$x = \lambda_D \sinh \omega_{pi} t. \qquad (6.31)$$

In comparison with Eq. (6.28) and assuming $eV_0 / kT_e \approx 10^4$, the maximum extraction time for an ion starting at the plasma boundary results as $\sim 5\omega_{pi}^{-1}$, so that the mean ion extraction time from the matrix sheath is approximately $2 \ldots 3 \cdot \omega_{pi}^{-1}$.

As so far neglected, the plasma boundary provides a continuous ion flux which is given by the Bohm flux. These ions are accelerated due the applied bias. At sufficiently large bias the initial kinetic energy can be neglected, so that the kinetic energy profile is given by the electrostatic potential in the sheath

$$\frac{m_i}{2} (v_i(x))^2 = -e\Phi(x). \qquad (6.32)$$

Continuity requires

$$e n_i(x) v_i(x) = j_B. \qquad (6.33)$$

This yields

$$n_i(x) = \frac{1}{e} j_B \left(-\frac{2e\Phi(x)}{m_i} \right)^{-1/2}. \tag{6.34}$$

The electron density in the sheath is neglected due to the strong repulsion by the negative bias. Then with the Poisson equation

$$\frac{d^2\Phi(x)}{dx^2} = -\frac{1}{\varepsilon_0} j_B \left(-\frac{2e\Phi(x)}{m_i} \right)^{-1/2}. \tag{6.35}$$

As above, we multiply with $d\Phi/dx$ and integrate. The result is

$$\frac{1}{2} \left(\frac{d\Phi(x)}{dx} \right)^2 = \frac{2}{\varepsilon_0} j_B \left(\frac{2e}{m_i} \right)^{-1/2} (-\Phi(x))^{1/2}. \tag{6.36}$$

Integration yields with $\Phi(x_s) = -V_0$

$$j_B = \frac{4}{9} \varepsilon_0 \left(\frac{2e}{m_i} \right)^{1/2} \frac{V_0^{3/2}}{x_s^2}. \tag{6.37}$$

This is the Child–Langmuir law of space-charge limited flux of charged particles. It confirms that the sheath thickness increases with the voltage, while the ion flux remains constant. The sheath thickness is after replacing j_B

$$x_s = \frac{\sqrt{2}}{3} \exp(1/2) \lambda_D \left(\frac{2eV_0}{kT_e} \right)^{3/4}. \tag{6.38}$$

As $eV_0 \gg kT_e$, the Langmuir–Child sheath is significantly thicker than the ion matrix sheath. Therefore, following the ion matrix extraction the sheath has to expand with the Child–Langmuir sheath as the stationary limit for long times.

The expansion phase is characterized by the so-called quasi-static Child-Langmuir shield, which means that the ion velocity in the sheath is fast compared to the sheath expansion velocity. During the expansion, a second influx into the sheath results as the plasma edge is "peeled off" by the expanding sheath, resulting in

$$n_e \exp(-1/2) \left(\sqrt{\frac{kT_e}{m_i}} + \frac{dx_s(t)}{dt} \right) = \frac{4}{9} \varepsilon_0 \left(\frac{2e}{m_i} \right)^{1/2} \frac{V_0^{3/2}}{(x_s(t))^2}. \tag{6.39}$$

Solutions of this differential equation can be found in literature. More rigorous treatments combine the matrix extraction with the quasi-static Child-Langmuir expansion. Results are shown in Fig. 6.4, which also shows the temporal dependence of the ion flux. The ion matrix extraction delivers a very strong flux peak. In agreement with the above consideration, it lasts

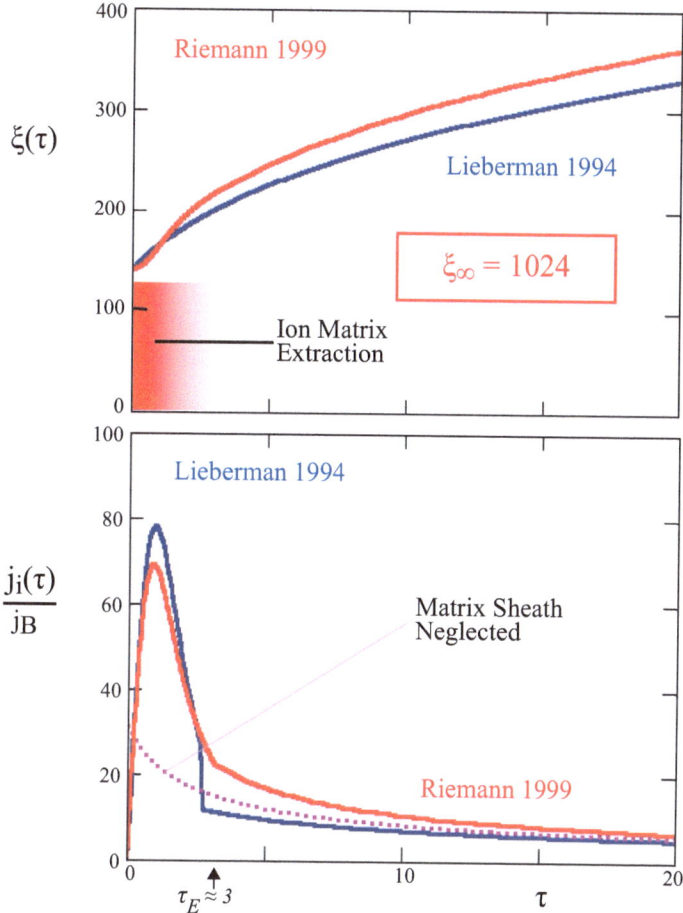

Figure 6.4: Sheath expansion (top) and ion flux to the electrode (bottom) during the ion matrix extraction and quasi-static Child-Langmuir phases, for an applied negative bias voltage of $10^4 k T_e/e$. The scaled dimensionless quantities are $\tau = \omega_{pi} t$ and $\xi = x_s/\lambda_D$. The ion flux is normalized to the Bohm flux so that the curves in the lower graph tend toward 1.

for about $3\omega_{pi}^{-1}$. After this, the ion flux remains still considerably higher than the stationary Bohm flux for a rather long time compared to the matrix extraction phase. The sheath thickness approaches the stationary value given by Eq. (6.38).

For the collision-free sheath, the ion energy at the electrode becomes

$$E_{i,b} = e(\Phi_{fl} + V_0), \qquad (6.40)$$

where Φ_{fl} can be neglected at sufficiently high bias.

6.5   COLLISIONAL SHEATH

At sufficiently high pressure, i.e., typically at pressures around 1 Pa or higher, collisions in the sheath have to be taken into account. This becomes necessary in particular in case of high bias, when the sheath thickness becomes large.

If no ions are generated or lost in the sheath, flux conservation (Eq. (6.33)) is still valid. Also, charge-transfer collisions do not change the number of ions, although they might remove the complete momentum from individual ions. However, the ions are no longer accelerated throughout the sheath, but attain a stationary drift velocity

$$v_i(x) = \mu_i E(x). \tag{6.41}$$

We first consider the case with a constant mean free path length throughout the sheath. Then, with Eq. (5.4),

$$\mu = \frac{e}{m_i \nu_c} = \frac{e\lambda_c}{m_i v_i(x)} \tag{6.42}$$

and

$$v_i(x) = \sqrt{\frac{e\lambda_c}{m_i} E(x)}. \tag{6.43}$$

With the same procedure as for Eq. (6.38) the result becomes

$$j_B = \frac{2}{3} \left(\frac{5}{3}\right)^{3/2} \varepsilon_0 \left(\frac{e\lambda_c}{m_i}\right)^{1/2} \frac{V_0^{3/2}}{x_s^{5/2}}. \tag{6.44}$$

Apart from a constant factor, this is different from the Child-Langmuir formula by the factor $(\lambda_c/x_s)^{1/2}$, which is small compared to one if an ion traveling through the sheath suffers many collisions. At constant flux, the sheath thickness becomes correspondingly smaller than in the Child-Langmuir collisionless case.

For the evaluation of the energy distribution of ions arriving at the negatively biased electrode, we consider the special case of charge transfer collisions according to Davis and Vanderslice. In particular in atomic gases such as Ar, symmetric charge transfer collisions exhibit a very large cross section (see Section 2.7).

In Fig. 6.5, where the potential at the electrode has been arbitrarily set to 0, ions starting at the plasma boundary are allowed to collide at any position x in the sheath. When collisions are dominant, the ions establish a drift velocity (Eq. (6.43)) so that the density is constant within most of the sheath. Therefore, the electric field and the electrostatic potential vary linearly and quadratically with the distance, respectively. With the correct boundary conditions,

$$x = x_s \left[1 - \sqrt{1 - \frac{V(x)}{V_0}}\right], \tag{6.45}$$

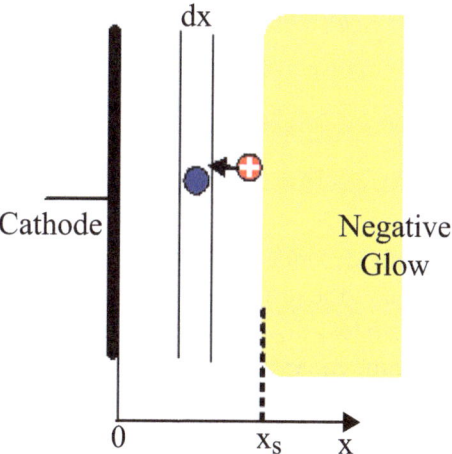

Figure 6.5: Cathode sheath geometry for the treatment of collisions of ions with neutral gas particles.

where $V_0 >> kT_e$ is the applied voltage.

Charge transfer collisions leave the flux of ions unchanged. Within a slab of thickness dx in the sheath, the number of charge transfer collisions per unit area is

$$d\dot{v}_{ct} = \frac{j_i}{\lambda_{ct}} dx, \tag{6.46}$$

where j_i is the ion flux and λ_{ct} the mean free path length between charge transfer collisions. The probability, that ions generated within dx reach the cathode without any further charge transfer collisions, is $\exp(-x/\lambda_{ct})$. Therefore, the flux of ions arriving from dx at the cathode is

$$dj_i(x) = \frac{j_i}{\lambda_{ct}} \exp\left(-\frac{x}{\lambda_{ct}}\right) dx. \tag{6.47}$$

(It should be noted that integration of Eq. (6.46) yields consistently the total ion flux provided the mean free path length is small compared to the sheath thickness.) Via Eq. (6.45), the start depth x corresponds to a potential difference $V(x)$. The ions resulting from charge transfer start with thermal energy close to zero, so that $V(x)$ corresponds to the ion energy $E_i = eV(x)$ at the cathode. The ion energy distribution at the cathode then results as

$$f(E_i) = \frac{1}{j_i} \frac{dj_i(E_i)}{dE_i} = \frac{1}{ej_i} \frac{dx}{dV} \frac{dj_i(x)}{dx}. \tag{6.48}$$

Inserting Eq. (6.47) and Eq. (6.45) in differentiated form, and changing to a normalized energy variable yields

$$f\left(\frac{E_i}{eV_0}\right) = \frac{x_s}{\lambda_{ct}}\left(1 - \frac{E_i}{eV_0}\right)^{-\frac{1}{2}} \exp\left(-\frac{x_s}{\lambda_{ct}}\left[1 - \left(1 - \frac{E_i}{eV_0}\right)^{\frac{1}{2}}\right]\right). \tag{6.49}$$

The shape of this distribution only depends on the ratio of the sheath thickness and the mean free path length. Figure 6.6 displays the result for different values of this parameter.

Depending on the mean path length, a fraction of the ions starting at the plasma boundary survives without any collisions, thus carrying the full energy eV_0. This fraction is given by

$$\alpha_0 = \exp\left(-\frac{x_s}{\lambda_{ct}}\right) \tag{6.50}$$

and has been added to Fig. 6.6 as a Gaussian distribution around $E_i = eV_0$, with an arbitrarily chosen width. For $x_s/\lambda_{ct} > 10$, the full-energy fraction becomes negligible.

It should be noted that fast neutrals have been neglected here. Each charge exchange collision creates a fast neutral. In terms of surface interaction, energetic neutrals behave similar to ions of the same energy, although they do not contribute to the electric current. Therefore, there neglect is not justified. However, when the mean free path length is sufficiently small, the ions gain little energy between the collisions, so that the energy of the neutrals is also small. Then, in addition, the neutrals lose energy in collisions with the gas atoms so that they quickly thermalize, so that any additional contribution to surface interaction can be neglected at sufficiently high pressure.

6.6   ELECTROSTATIC PROBE

An electrostatic or Langmuir probe (Fig. 6.7) is a small electrode which is introduced into the plasma for simple diagnostics. A bipolar DC power supply is connected which delivers a voltage V. If the voltage is sufficiently negative, it draws an ion current which is given by Eq. (6.21) and which, in the present context, is called the "ion saturation" current (for simplicity, we assume here a flat probe.) Increasing the voltage to the floating potential V_{fl}, the ion saturation current is compensated by an increasing electron current, which is derived from the local electron density according to the Boltzmann relation, Eq. (4.30). Thus, above the floating potential, electrons are less and less repelled by the electrode, until finally the full flux of electrons is delivered to the electrode when V becomes equal to the plasma potential V_{pl}. Then, the plasma boundary is significantly disturbed, with the ions now being repelled at increasing voltage. As, however, the ion saturation current is much smaller than the electron saturation current, one may, in a first

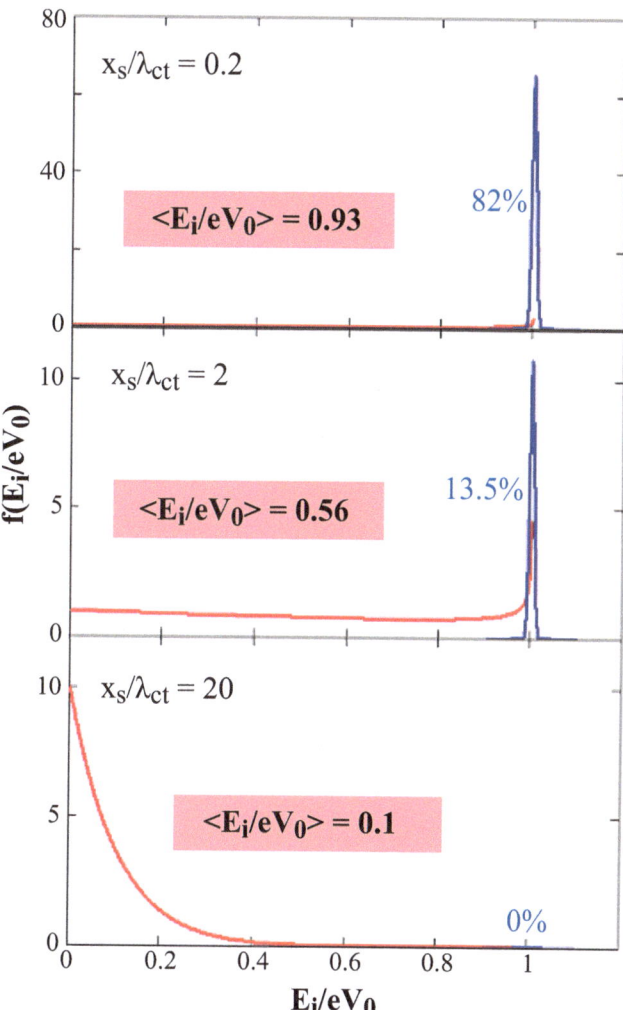

Figure 6.6: Ion energy distributions resulting from charge transfer collisions according to Davis and Vanderslice (red lines) and from the full-energy fraction (blue lines), for different ratios of sheath thickness and mean free path length. The full-energy peaks have been arbitrarily assumed as Gaussian with a standard deviation of $5 \cdot 10^{-3}$. The full-energy fractions and the mean relative energies are indicated.

approximation, neglect the decrease of the ion current and write the total current as

$$I(V) = I_i + I_e(V) = Aen_e\sqrt{\frac{kT_e}{m_i}}\exp(-1/2) + \frac{Aen_e}{4}\sqrt{\frac{8kT_e}{\pi m_e}}$$

$$\times \begin{cases} 1 & \text{if } V > V_{pl} \\ \exp\left(\frac{e(V-V_{pl})}{kT_e}\right) & \text{else} \end{cases}$$

(6.51)

in accordance with Section 6.3 and with A denoting the probe area. The resulting curve is qualitatively shown in Fig. 6.8. In practice, the measured current in the electron saturation regime is not constant due to the small area of the electrode. Toward high positive voltages, the thermal motion of the electrons around the probe is influenced by the electric field, and electrons are drawn to the probe from an increasing volume around it. Therefore, above the plasma potential the current often increases, which is indicated schematically.

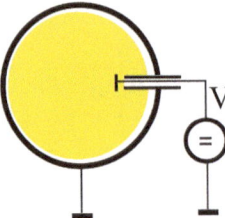

Figure 6.7: Langmuir probe arrangement for plasma diagnostics, with a bipolar DC voltage supply.

More rigorous treatments are available in the literature. The electron retardation regime below V_{pl} carries information about local the electron energy distribution function of the plasma, from which in particular the electron temperature can be derived. The electron density is calculated from the saturation currents.

As a remark of caution, the floating potential V_{fl} delivered by the Langmuir probe measurement is generally different from the floating potential Φ_{fl} to which an electrodeless plasma adjusts in contact with a wall (see Section 6.3, Eq. (6.25)). The probe measures local plasma and plasma boundary properties which, due to inhomogeneities of the charged particle distribution, are in general different from the properties at the chamber walls. On the other hand, V_{pl} represents the local electrostatic potential of the plasma. In a homogeneous electrodeless plasma with a conducting wall, the plasma potential obtained from the probe diagnostics, V_{pl}, if measured with respect to the wall, is identical to the floating potential of the plasma, Φ_{fl}.

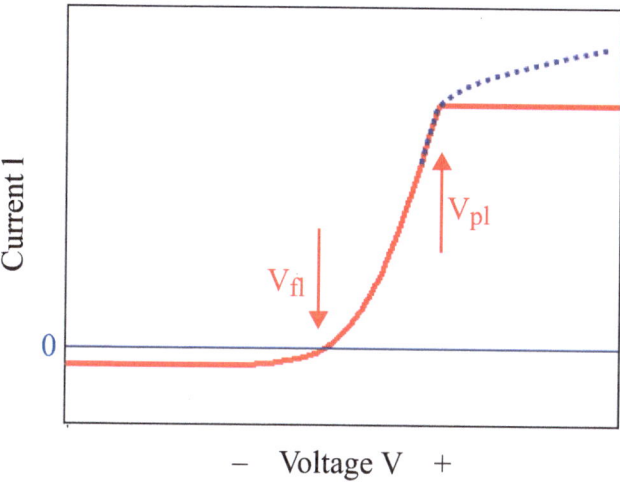

Figure 6.8: Current-voltage characteristics of a Langmuir probe (qualitative), the (negative) ion branch is artificially enhanced vs. the electron branch. The red line results from the idealized Eq. (6.51), whereas the blue dotted line indicates a more realistic behavior. The floating and plasma potentials are indicated.

CHAPTER 7

Plasma-surface Interaction

The interaction of the plasma with the surrounding solid surfaces, i.e., the interaction of thermal and energetic particles from the plasma with the atoms of the solid, is a wide topic which cannot be treated here in detail. Also, due to the wide spectrum of species being generated in a plasma, in particular in a reactive plasma, the complete description of the plasma-wall interaction is often an insolvable task, which is even complicated by he fact that ions at atoms might be in excited states when hitting the wall. Therefore, in many areas the understanding of plasma-wall interaction is presently still at its infancy.

In principle, the bulk plasma and the plasma-wall interaction cannot be treated independently. Species lost from the plasma interact with the wall, but this interaction may produce particles which are reemitted into the plasma.

In the following, we will concentrate on the main mechanisms of plasma-wall interaction which are essential for plasma physics and processing. For brevity, we will refrain from any derivations and just display the findings from literature.

7.1   ION IMPLANTATION AND REEMISSION

A fast ion impinging onto a solid may enters a subsurface region where it looses its energy by collisions with the atomic cores and the electrons of the solid. The atomic core interaction may be treated as subsequent colllisions in the screened Coulomb potential of the atomic nuclei. Between and during the nuclear collisions, electronic collisions occur with the atomic electrons or the electrons of the free electron gas of the solid. This is schematically displayed on the right-hand side of Fig. 7.1.

The energy loss of the fast ion in the solid per unit path length, dE/ds, is given by the so-called stopping cross section S, which is the energy loss per transversed atomic areal density of the solid, and the atomic density of the solid:

$$\frac{dE}{ds} = nS. \tag{7.1}$$

Electronic and nuclear interaction are treated independently so that the stopping is divided into an electronic and a nuclear fraction

$$S = S_e + S_n. \tag{7.2}$$

From most plasmas, the walls are bombarded with ions the velocity of which is small compared to the mean velocity of the electrons of the solid. In this regime, the electronic stopping

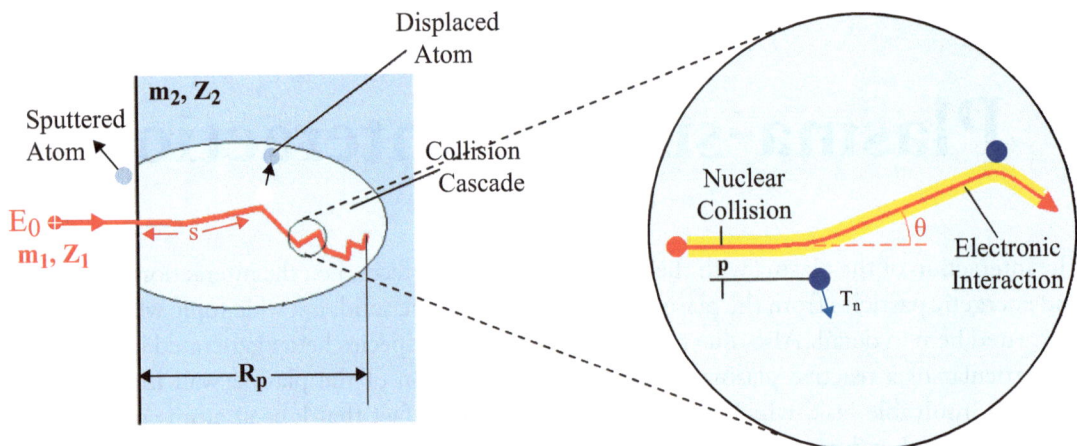

Figure 7.1: Collisional phenomena of ion-solid interaction: ion range, sputtering, and radiation damage (left), and nuclear collisions and electronic interaction (right).

is proportional to the ion velocity. An approximate universal formula reads

$$S_e(v) = \frac{8\pi e^2 a_0}{4\pi \varepsilon_0} \frac{Z_1^{7/6} Z_2}{(Z_1^{2/3} + Z_2^{2/3})^{3/2}} \frac{v}{v_0}, \tag{7.3}$$

where a_0 and v_0 denote the radius and the velocity, respectively, of the first Bohr orbit, and Z_1 and the Z_2 the atomic numbers of the ion and the solid atoms, respectively. The nuclear stopping is obtained from the ion scattering in a screened Coulomb potential. An approximate universal fit formula is

$$\left(-\frac{d\varepsilon}{d\rho}\right)_n = \frac{3.44\sqrt{\varepsilon}\log(\varepsilon + 2.718)}{1 + 6.35\sqrt{\varepsilon} + \varepsilon(6.882\sqrt{\varepsilon} - 1.708)}, \tag{7.4}$$

where ε and ρ are reduced dimensionless variables for the ion energy and the path length, respectively,

$$\varepsilon = \frac{a}{b} = \frac{4\pi \varepsilon_0 a}{Z_1 Z_2 e^2} \frac{m_2}{m_1 + m_2} E, \tag{7.5}$$

where m_1 and m_2 represent the mass of the ion and the solid atoms, respectively, and

$$\rho = \pi a^2 n \frac{4 m_1 m_2}{(m_1 + m_2)^2} \cdot s \tag{7.6}$$

with a denoting the screening lenght in the Thomas–Fermi model

$$a = \frac{0.8853 a_0}{(Z_1^{2/3} + Z_2^{2/3})^{1/2}}. \tag{7.7}$$

Examples are given in Fig. 7.2 for hydrogen and argon ions in silicon. The stopping increases strongly with the atomic number of the ion. For light ions, electronic stopping is dominant above ~ 100 eV, whereas the stopping of heavy ions is dominated by nuclear, stopping over the entire present energy range.

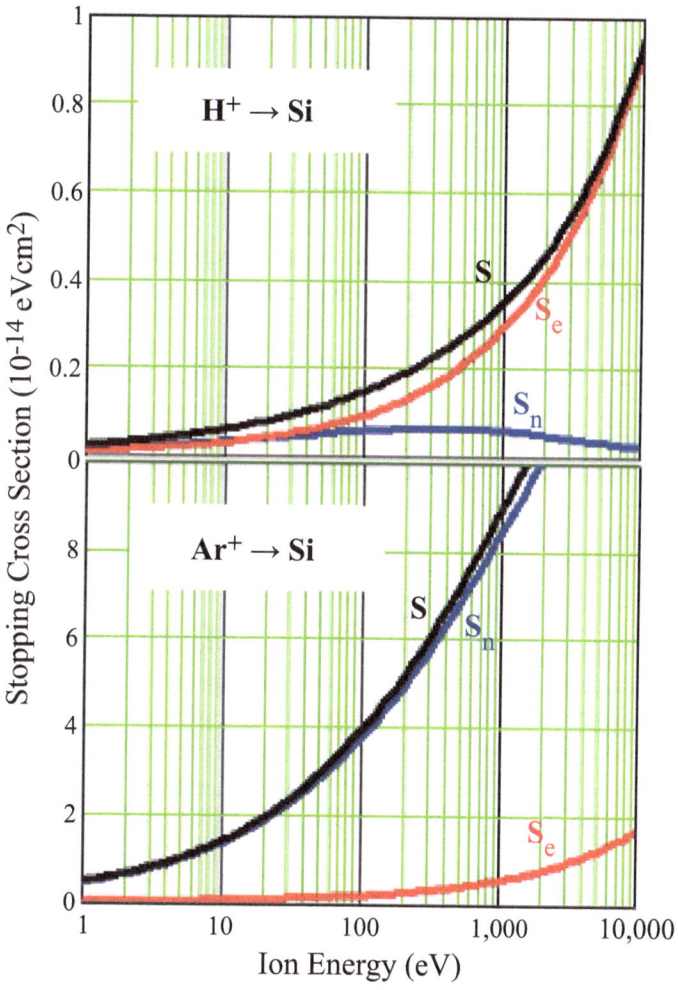

Figure 7.2: Electronic, nuclear, and total stopping of hydrogen and argon ions in silicon.

Semi-empirical stopping data are well documented in literature. From the stopping cross section, the total path length of an ion is directly obtained by integration

$$\bar{R}_t(E) = \int_0^{\bar{R}_t} ds = \frac{1}{n} \int_0^E \frac{dE'}{S(E')}. \tag{7.8}$$

However, as indicated in Fig. 7.1, the total path length along the trajectory can be substantially larger than the ion implantation depth, which is given ba the mean projected range R_p. Only for ions with masses large compared to the target mass, where the angular deflections during the scattering remain small, the total path length is a good approximation for the projected range. In general, transport calculations of the latter are complicated. Therefore, nowadays computer simulations of statistically distributed ion trajectories are preferred to obtain the mean projected range and its depth distribution. The freely available SRIM program package offers both computer simulations and analytical range algorithms. Figure 7.3 shows examples of ion penetration depths over a wide range of incident energies. A rule-of-thumb is that 1 keV incident energy corresponds to a mean projected range of 1 nm for heavy ions and 10 nm for the lightest ions.

In Fig. 7.4, ion implantation profiles are shown which have been generated by computer simulation. At energies being characteristic for plasma-wall interaction, the distributions are rather broad.

Toward lower energies, there is also an increasing probability, in particular for light ions, that the ion momentum is reversed by a sequence of nuclear collisions, so that the ion leaves the surface again before being stopped (see Fig. 7.5).

The reflection coefficient r is defined as the ratio of the flux j_r, which is reemitted into the plasma due to this kinematic reflection, and the incident ion flux j_i,

$$r = \frac{j_r}{j_i}. \tag{7.9}$$

The energy spectrum of the reflected atoms is broad and extends to close to the incident energy. Furthermore, there is a certain probability that the ion captures an electron from the solid before being reemitted. Restricting ourselves to low energies where the interaction with the ion is mainly restricted to surface atoms, the neutralization probability P_n is given according to the Hagstrum model by the time which is spent by the ion when back-crossing the surface, i.e., the normal velocity component for the outgoing trajectory. The model yields

$$P_n = 1 - \exp\left(-\frac{v^*}{v_\perp}\right), \tag{7.10}$$

where v^* denotes a velocity which depends on the electronic transition rates and screening. As $v_\perp$ scales with the incident energy, the neutralization probability will increase at decreasing incident ion energy. However, a quantitative application of Eq. (7.10) is questionable.

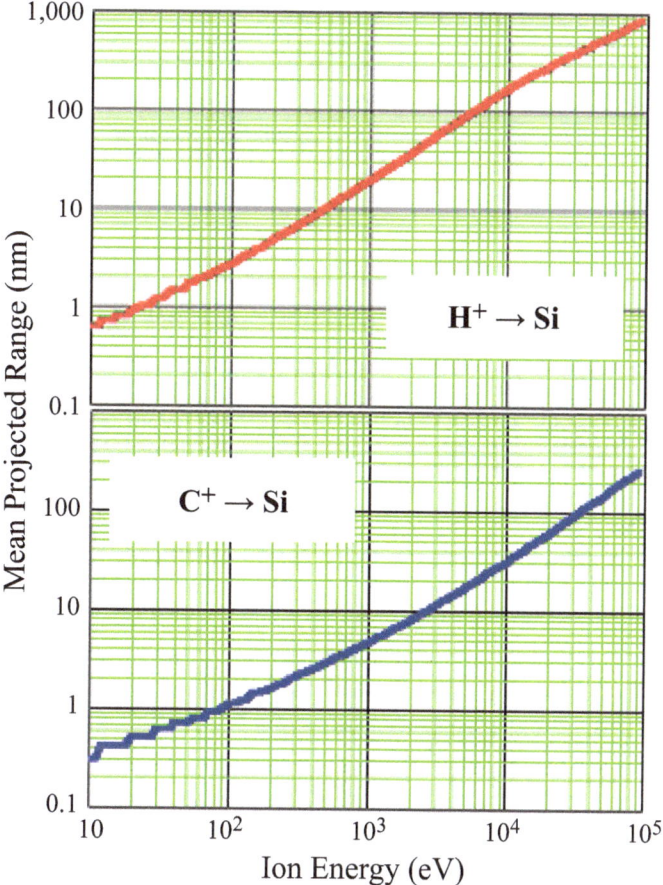

Figure 7.3: Mean projected ion ranges versus the implantation energy for protons and carbon ions in silicon. The data are obtained from the PRAL range algorithm which is available in the SRIM program package (3H http://www.srim.org).

When the ion energy becomes very low such as resulting with a nonbiased wall from the floating potential only, i.e., without additional bias, the ion penetration depth falls below 1 nm. Then, the above concept of ion implantation becomes questionable. The ion may no longer be able to "squeeze" through the top atomic layer. This is schematically indicated in Fig. 7.6.

The probability of subsurface implantation (sometimes also called subplantation) can be calculated by means of computer simulation. An example is given in Fig. 7.7. At an energy above 100 eV, all ions are implanted as the kinematic reflection is negligible for ions with equal or higher mass than the target atom mass. Below 20 eV, the surface penetration probability is zero. This is a rather general finding with changes of the characteristic energies for differ-

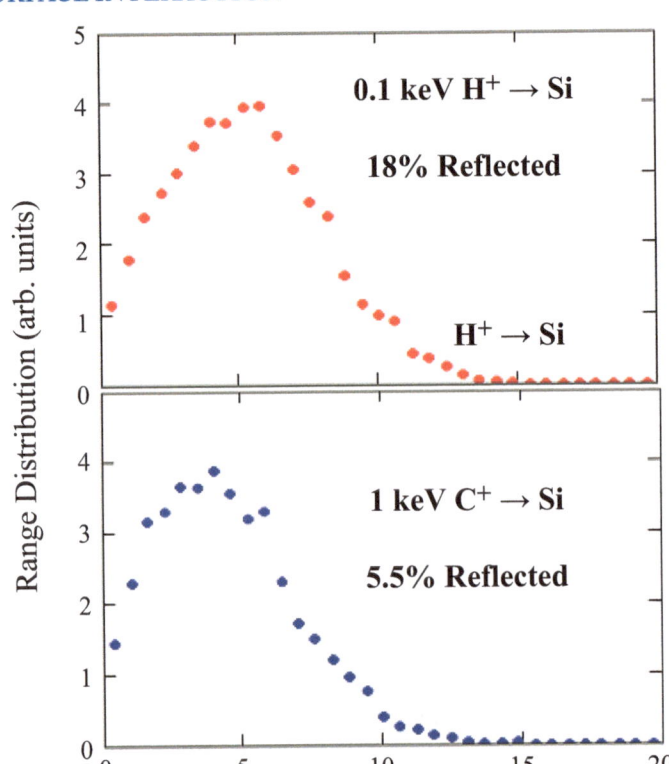

Figure 7.4: Mean projected ion ranges for protons and carbon ions in silicon for the specified implantation energies. Also, the reflection coefficients are indicated. The data have been generated by TRIM binary collision computer simulation which is available in the SRIM program package (2H http://www.srim.org).

ent systems which are smaller than about an order of magnitude. This indicates that a floating low-temperature plasma will only to some extend provide ions which get implanted. A sizeable fraction of ions will be deposited at the surface, and thus may be treated like thermal neutrals.

So far, we have only addressed collisional phenomena. However, when the ion comes to rest, it will finally thermalize with the atoms of the solid and be subjected to the physics and chemistry of the solid. This applies to both subplanted ions and to those which reside at the surface after an unsuccessful attempt of surface penetration. Ions which are reactive with the atoms of the solid, will then normally stay at the position to which they have been slowed down. In contrast, inert ions may desorb from the surface either directly or after diffusion out of the subsurface layer. These phenomena are strongly influenced by the temperature of the solid and

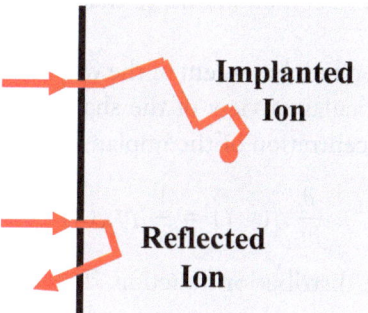

Figure 7.5: Ion implantation and reflection (schematic).

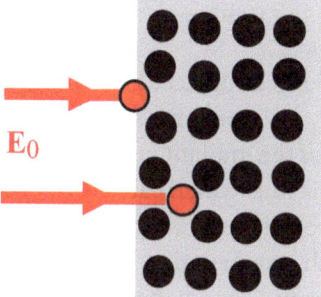

Figure 7.6: An ion with an incident energy in the surface penetration threshold regime may either be implanted below the surface (bottom) or reside at the surface (top) provided it is reactive.

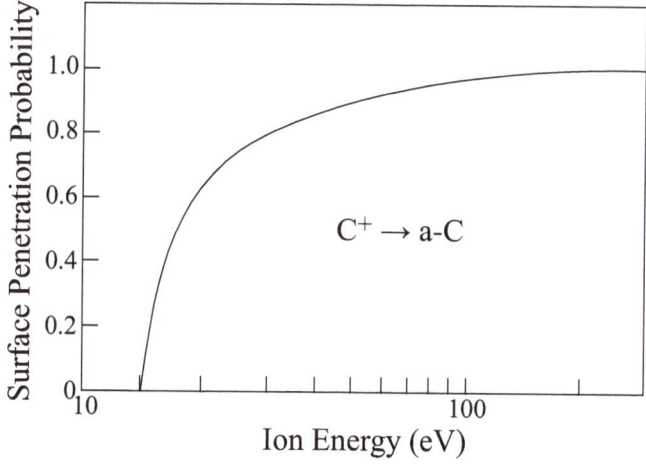

Figure 7.7: Surface penetration probability of carbon ions incident on amorphous carbon as obtained from TRIM binary collision computer simulation.

differ widely for different ion-solid combinations, so that any general quantitative prediction is excluded.

As there is a continuous ion bombardment of the wall, the concentration of the implanted atoms increases quickly, in particular in view of the shallow implantation depths. Neglecting thermal release, the volume concentration of the implanted atoms changes according to

$$\frac{\partial}{\partial t} c_i(x, t) = \frac{1}{n} j_i \, f_R(x), \qquad (7.11)$$

where $f_R(x)$ denotes the range distribution function. The simplest approach to high-fluence implantation profiles is to neglect any changes in n and f_R, which arise from the presence of the implanted species, and thereby just to scale the range distribution, so that

$$c_i(x, \Phi) = \frac{1}{n} f_R(x) \cdot \Phi, \qquad (7.12)$$

where $\Phi = j_i \cdot t$ denotes the implanted fluence. Equation (5.7) neglects any relaxation of the target substance and is thus strictly valid only for small relative concentrations.

In reality, the concentration of the implant in a solid is often limited, such as by a maximum concentration of implanted gaseous ions which can be accommodated, or stoichiometric limits when reactive ions are implanted forming a compound. This can be accounted for in the simple model of "local saturation," which assumes that any atom which is implanted into a region where the maximum concentration has already been reached, is immediately released from the substance. In this model, the profile evolution with a maximum concentration $c_{i,max}$ is given by

$$c_i(x, \Phi) = \begin{cases} \frac{1}{n} f_R(x) \cdot \Phi & \text{if } c_i < c_{i,max} \\ c_{i,max} & \text{else} \end{cases}. \qquad (7.13)$$

This is displayed schematically in Fig. 7.8. Above a certain critical fluence, reemission due to saturation starts. In the limit of high fluences, a rectangular implantation profile is established in a depth given by the maximum ion range, which is close to the mean integrated path length of Eq. (7.8).

In a low-pressure plasma, typical ion fluxes are around 10^{15} cm^{-2}s^{-1}. Let us assume a negative bias of a few 100 V leading to an implantation depth of a few nm (see Fig. 7.3) which corresponds to an areal density in the order of $10^{16} \dots 10^{17}$ atom/cm^2. For an implantation reaching stoichiometry within the ion range, a fluence of about half of this value is required. Thus, the stationary implantation profile is reached within a few to hundred seconds, so that a full reemission of the ionic species is established very quickly, unless the excess atoms diffusive into the depth or to the surface and are bonded there.

In summary, in stationary state a small fraction of the impinging ions is kinematically reflected into the plasma. The remaining, large fraction is either integrated into the wall, in particular when the ions are reactive, or fully reemitted thermally. In the latter case, the reemitted atoms are neutral, so that the total backflow into the plasma consists almost entirely of neutrals.

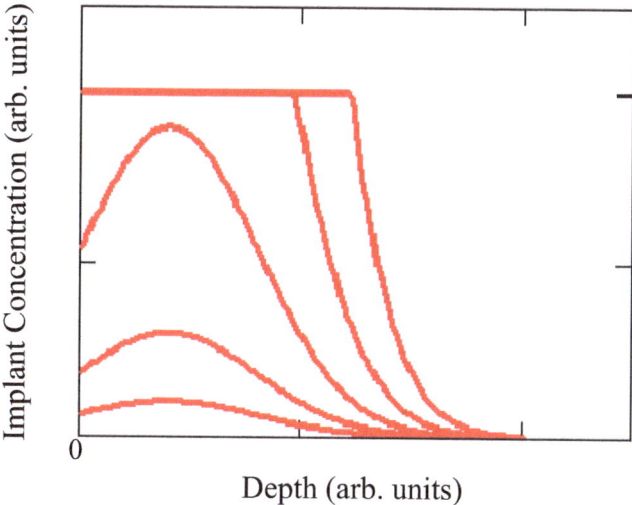

Figure 7.8: Model of local saturation (schematic). A Gaussian range distribution has been assumed with five ion fluences, incremented by a factor of 3 from one to another.

7.2   COLLISION CASCADE

As already indicated in Fig. 7.1 and detailed in Fig. 7.9, ions of sufficient energy may displace atoms of the solid from their lattice sites by elastic nuclear knock-on processes. The primary knock-on atoms may receive sufficient energy transfers (see Eq. (2.11)) to promote the formation of further branches of the collision cascade. It can be shown that the energy transfers T in the total cascade for an ion with an incident energy E_0 are distributed according to

$$f_c(T) \sim \frac{E_0}{T^2} \tag{7.14}$$

which indicates that many low-energy cascade atoms result from the primary ion. As long as the average kinetic energy of the moving atoms is larger than the atomic binding energy in the lattice, which is about 3 eV, the cascade is denoted as collisional and can well be described by two-body collisions. The collisional phase of the cascade is terminated after about 100 fs. During the following further "hyperthermal" phase, the cascade atoms are thermalized toward the temperature of the solid, so that it has to be modeled by many-body collisions. A local temperature of ~ 0.1 eV is reached within $\sim$ ps to about 10 ns, depending on the thermal properties of the solid.

Below, we will concentrate on the collisional phase of the cascade. For its lateral extension a few nm are a good, so that a projected area of about 10^{-13} cm^2 is affected by a single ion.

Assuming an ion flux of $\sim 10^{16}$ cm^{-2}s^{-1} which is typical for dense low-temperature plasmas, about 103 ions hit this area per s. As the lifetime of the collisional phase is only $\sim 10^{-13}$ s,

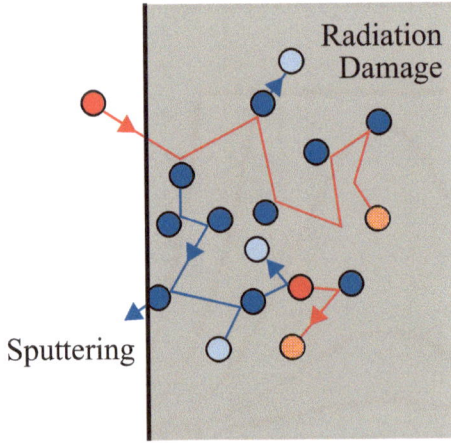

Figure 7.9: Collision cascade initiated by an incident ion. Cascade atoms at the surface may contribute to sputtering, atoms in the bulk to radiation damage.

the probability that the collisional phases of two subsequent ions overlap is $\sim 10^{-10}$. Thus, the nonlinear superposition of cascades can be neglected even at ion fluxes being orders of magnitude higher.

7.3   RADIATION DAMAGE

Recoil atoms in the cascade are able to displace lattice atoms permanently from their original lattice sites provided the energy transfer exceeds an average displacement threshold energy, U_d, which lies between 20 eV and 80 eV for different solids. Such an event creates a "Frenkel pair" which consists of a vacancy at the original site and an interstitial atom at the new position of the recoil. The number of Frenkel, N_{FP}, per incident pairs can be calculated with sufficient precision by modeling the cascade atoms as hard spheres. The result is the Kinchin-Pease formula

$$N_{FP}(E_0) \approx \begin{cases} 0 & \text{if } E_0 < U_d \\ \frac{E_0}{2U_d} & \text{else} \end{cases} . \tag{7.15}$$

Therefore, under floating conditions, the number of produced Frenkel pairs per incident ions is rather small and often vanishes. On the other hand, biased substrates may receive a significant damage during plasma processing. We will come back to this in a later chapter.

At elevated temperatures, interstitials and vacancies become mobile. In metals, interstitials become mobile already below 50 K, vacancies significantly above room temperature. In elemental semiconductors, interstitials and vacancies are stable to slightly below room temperature. When becoming mobile, the point defects might annihilate, cluster into agglomerates such as dislocation loops or voids, or migrate to sinks such as grain boundaries or the surface. At a given temperature, these phenomena are widely different for different materials. When no bias is applied, the Frenkel defects are generated very close to the surface so that their concentration remains very small in case of thermal mobility.

7.4   SPUTTERING

A cascade atom can be transmitted through the surface if its initial energy exceeds the surface binding energy. For the surface binding energy, the sublimation enthalpy U_s of the solid is a good approximation, with values between 2 eV and 8 eV for elemental solids. The sputtering yield Y_s is defined as the number of sputtered atoms per unit area and time, j_s, divided by the incident ion flux

$$Y_s = \frac{j_s}{j_i}. \tag{7.16}$$

From transport theory, the sputtering yield is obtained according to the Sigmund formula

$$Y_s(E_0) = \frac{4.2 \cdot 10^{14}\ \mathrm{cm}^2}{U_s} \cdot \zeta\left(\frac{m_2}{m_1}\right) \cdot S_n(E_0). \tag{7.17}$$

$\zeta(m_2/m_1)$ varies from 0.15–0.8 as function of the target-to-ion mass ratio. Eq. (7.17) demonstrates that the collisional sputtering yield follows the nuclear stopping. Fig. (7.10) shows the Sigmund results for a special ion-solid combination.

For light ions, the necessary momentum reversal can be conveniently pictured as shown in Fig. 7.11. The ion is backscattered from a target atom, then hits a surface atom, which is then ejected. If 180° backscattering is assumed for both collisions, the energy of the surface recoil becomes according to Eq. (2.11)

$$E_{rs} = \gamma(1-\gamma)E_0 \tag{7.18}$$

when electronic stopping is neglected. From this, sputtering threshold energy results according to

$$E_{s,th} = \frac{E_0}{\gamma(1-\gamma)}. \tag{7.19}$$

However, it has to be noted that a lower threshold results when assuming multiple scattering for the momentum reversal rather than a single collision, which prevents from a simple definition of the threshold energy for sputtering.

The threshold is neglected in Eq. (7.17). Consequently, semi-empirical modifications of the Sigmund formula have been presented in literature, as also shown in Fig. 7.10. It is further demonstrated that also binary collision computer simulations can be reliably employed to predict sputtering yields.

Sputtering is almost negligible for a floating plasma. At negative substrate bias of 100 V and more, however, the sputtering yields become sizeable.

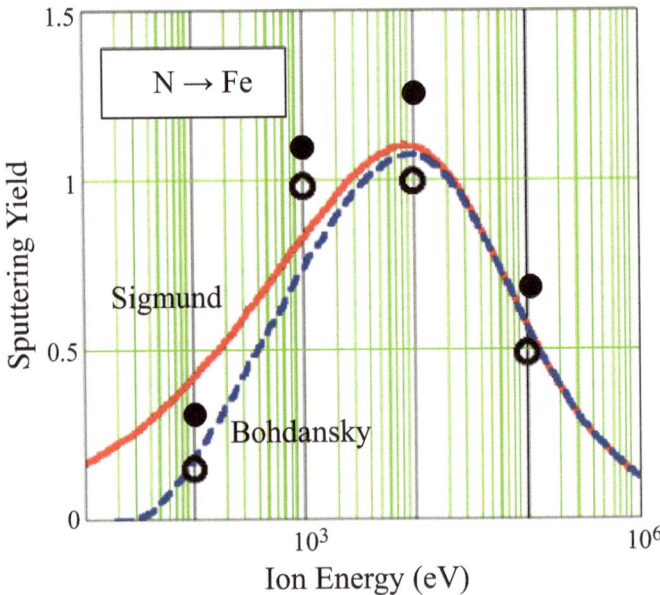

Figure 7.10: Sputtering yield versus ion energy for nitrogen ions incident on iron. The red line has been calculated from Eq. (7.17). Also shown are a corrected prediction which accounts for threshold effects (blue dashed line) and the results from different binary collision computer simulations (dots).

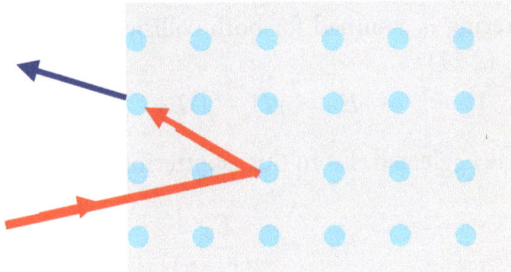

Figure 7.11: Simple mechanism of light ion sputtering demonstrating the existence of a sputtering threshold energy.

For most solids, the emitted sputtered atoms are neutralized to almost 100%, due to their typically low energies. The energy of the sputtered atoms, E_s, is distributed according to the Thompson formula

$$f_E(E_s) \sim \frac{E_s}{(E_s + U_s)^3} \tag{7.20}$$

which is shown in Fig. 7.12. The function peaks at an energy which is given by $U_s/2$ and has a long high-energy tail. The mean energy of the sputtered particles at a given incident energy is given by

$$\bar{E}_s \approx U_s \left[2 \log \left(\gamma \cdot \zeta \left(\frac{m_2}{m_1} \right) \cdot \frac{E_0}{U_s} \right) - 3 \right]. \tag{7.21}$$

For a surface binding energy of 5 eV and an incident ion energy of 500 eV, this yields for $m_1 = m_2$ a mean energy of 15 eV.

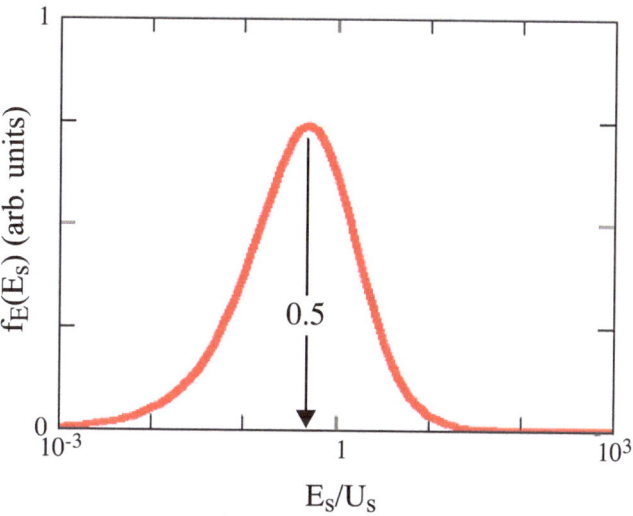

Figure 7.12: Energy distribution of the sputtered atoms according to the Thompson formula.

7.5   CHEMICAL SPUTTERING

For light ions, Eq. (2.11) reduces to $\gamma = 4 m_1 / m_2$. For the example of hydrogen impinging on a solid with a medium atomic number, $\gamma \approx 0.05$ resulting in a threshold energy of around 100 eV according to Eq. (7.19), below which the sputtering yield becomes negligible. In this regime, effects of hot surface chemistry may enhance the emission surface atoms, in particular if reactive ions like hydrogen are impinging.

In connection with the plasma wall interaction in nuclear fusion devices, the chemical sputtering has been investigated to some detail in for hydrogen ions interacting with graphite. Several mechanisms have been identified. Continuous implantation of hydrogen builds up a hydrogenated surface layer according to the model of local saturation (see Section 7.1) with a hydrogen concentration of about 30% up to temperatures of about 300°C. In this layer, excess hydrogen and radiation damage may result in the formation of a CH_4 molecule which is mobile and can be released through the surface. This is indicated in Fig. 7.13. The effect is strongly dependent on temperature, as the molecule formation and its migration are thermally activated. Toward higher temperature, it is limited by the decreasing hydrogen concentration in the implanted layer. This results in a maximum around 600°C with a maximum yield of C emission of about 0.1 for 1 keV hydrogen ions. Compared to this, the physical sputtering yield according to the Bohdansky formula (see Fig. 7.10) is about 0.02.

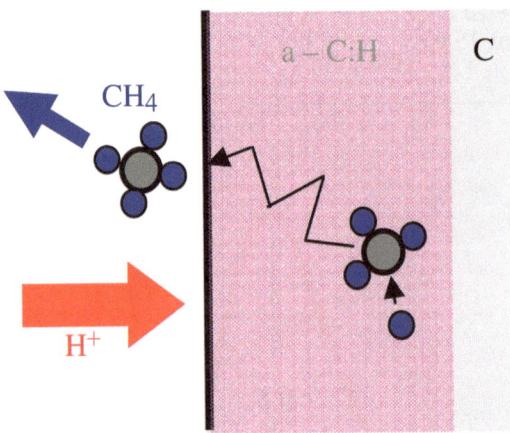

Figure 7.13: A mechanism of chemical sputtering by internal molecule formation, diffusion, and surface release. The hydrogen implantation results in an amorphous hydrogenated carbon layer.

Another mechanism has been demonstrated recently by molecular dynamics simulations, as shown in Fig. 7.14. A hydrogen ion with an energy between 10 and 30 eV, which is below the threshold energy of physical sputtering, induces an abstraction of a methyl radical from the surface of graphite. This process takes place at room temperature. The yield increases with the hydrogen ion energy from about 0.005–0.05.

Summarizing, chemical sputtering of carbon delivers small yields, but can play an important role as it introduces new chemical species into the plasma. Very little is known about chemical sputtering for other ion-solid combinations.

7.6   SURFACE REACTIONS

Chemical reactions at the surface cannot only be induced by fast ions emitted from a plasma, but also by reactive neutrals which are produced in a plasma of reactive gases. In general, there are only a few conclusive informations in this area, as the surface reactions are very difficult to access experimentally. On the other hand, their understanding is important as their products may contribute essentially to the composition of the plasma.

A selection of possible surface processes induced by ions, hydrogen atoms, and radicals is depicted schematically in Fig. 7.15. Quantitative descriptions of such processes are scarce and available only for comparatively simple systems such as methane or silane plasmas. The sequence

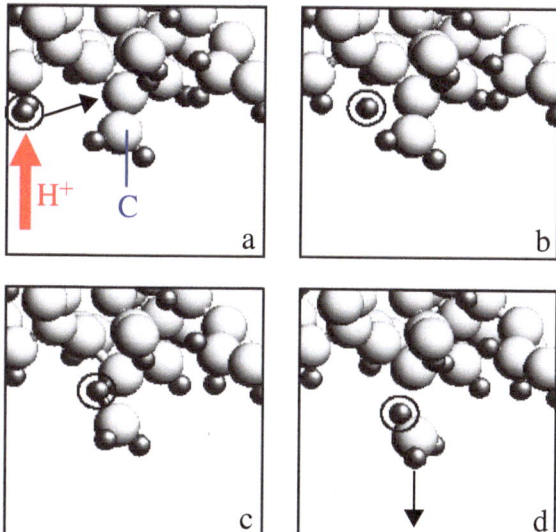

Figure 7.14: Molecular dynamics simulation of the the room temperature chemical sputtering of amorphous carbon by hydrogen bombardment. The ion hits a surface atom (a) and is scattered sideways (b). It inserts between a carbon-carbon bond (c) and causes the release of a CH_3 radical (d).

and interaction of the individual processes determines the plasma-wall chemistry, which, e.g., may result in surface activation, surface etching, or thin film deposition by means of reactive plasmas.

For the purpose of illustration, we will select here a simple example for the plasma-induced polymerization from a low-pressure methane plasma, using the so-called "adsorbed layer" model. As shown in Fig. 2.13, electron impact produces CH_4^+ and CH_3^+ ions, as well as CH_3 and H^0 radicals in the plasma. From these, the almost identical ions will be subsumed as one ionic species. The hydrogen is neglected for the present purpose, so that only the processes (d) and (e) of Fig. 7.15 will be considered, i.e., the adsorption of methyl radicals and the ion-induced integration of adsorbed radicals into the growing film. In addition, thermal desorption of adsorbed radicals is taken into account. The mean residence time of the adsorbate is given by

$$\tau_d = \tau_0 \exp\left(\frac{U_d}{kT_s}\right), \tag{7.22}$$

where $\tau_0 \approx 10^{-12}$s denotes the inverse of the attempt frequency of desorption, U_d the sorption energy, and T_s the substrate temperature. Let us assume that the surface contains ν_0 physisorption sites per unit area. The surface coverage θ_s denotes the fraction of these sites which are occupied. With the sticking coefficient of the radicals S_0, the ion flux j_i, the radical flux j_n, and

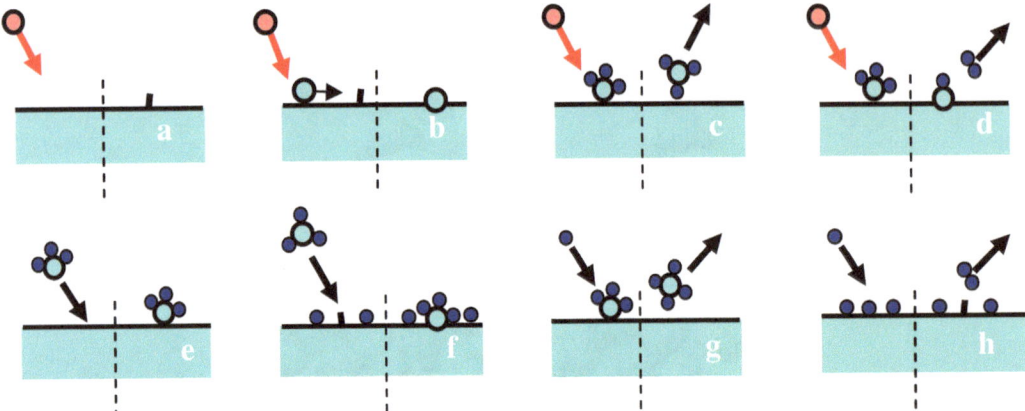

Figure 7.15: Surface processes induced by ions (top) and neutral radicals (bottom). (a) Ion-induced surface damage producing, e.g., a dangling bond. (b) Ion-induced mobility and surface insertion of an adsorbed atom. (c) Ion-induced desorption of adsorbed species. (d) Ion stitching causing surface insertion of an adsorbed atom, accompanied by hydrogen molecule emission. (e) Physisorption of a radical. (f) Insertion of a radical at a dangling bond. (g) Abstraction of a physisorbed radical by atomic hydrogen. (h) Hydrogen abstraction by atomic hydrogen, creating a dangling bond.

the ion stitching cross section σ_{st} the balance equation for the coverage reads in stationary state

$$\frac{d\theta}{dt} = \frac{j_n S_0}{v_0}(1 - \theta) - \frac{\theta}{\tau_d} - \theta j_i \sigma_{st} = 0. \tag{7.23}$$

From this, the coverage can be calculated. The film growth rate (increase in carbon areal density per unit of time) results then as

$$j_C = j_i + \theta v_0 j_i \sigma_{st} = j_i + \frac{j_n S_0 j_i v_0 \sigma_{st}}{j_n S_0 + v_0 \tau_d^{-1} + v_0 j_i \sigma_{st}}, \tag{7.24}$$

where the ion implantation has been included assuming zero reflection. Assuming further the low-temperature limit, that is infinite residence time of the adsorbate, and $\sigma_{st} \approx v_0^{-1}$, the deposition rate is controlled by ion implantation and adsorption if $j_i \gg j_n S_0$, and by ion implantation and ion stitching $j_i \ll j_n S_0$.

It is also interesting to calculate the effective sticking coefficient of the neutral radicals in the frame of the adsorbed layer model, which is defined by the difference of adsorbed and desorbed fluxes,

$$j_n S_{eff} = j_n S_0(1 - \theta) - \frac{v_0}{\tau_d}\theta \tag{7.25}$$

and results as

$$S_{eff} = \frac{j_i v_0 \sigma_{st}}{j_n S_0 + v_0 \tau_d^{-1} + v_0 j_i \sigma_{st}}. \tag{7.26}$$

This demonstrates that the effective sticking of the neutrals does not only depend on the temperature, but also on the ion flux. Thus, the transport of ions and neutrals can no longer be treated independently when surface reactions are included.

The area of surface chemistry in plasma processing has been very little investigated so far, so that a reliable modeling of reactive plasmas is hampered by the poor knowledge of surface processes.

7.7 SECONDARY ELECTRON EMISSION

Upon ion impact on a solid surface, the electronic interaction may create free electrons which are backtransmitted through the surface, the so-called secondary electrons. In analogy to the sputtering yield, the secondary electron emission coefficient γ_e (often called SEEC) is defined as the flux of emitted electrons, j_{se}, relative to the incident ion flux,

$$\gamma_e = \frac{j_{se}}{j_i}. \tag{7.27}$$

Secondary electron emission occurs by a variety of mechanism, which are classified as potential emission and kinetic emission.

In potential emission, the driving process is transfer of an electron from a surface atom when a slow ion approaches a surface, similarly as demonstrated for the charge transfer collisions in Section 2.6. Two selected atomic processes are shown in Fig. 7.16. During Auger neutralization, an electron tunnels from an electronic state to the ground atomic state of the ion. If the ionization energy of the atom, E_i, is larger than 2Φ where Φ denotes the work function of the solid, an electron can be emitted from the solid into the vacuum. During resonance neutralization an electron below the Fermi level tunnels into an empty state of the ions. An electron from the Fermi edge can fill the vacancy and transfer the released energy to another electron which is emitted. Again, the condition is that the ionization energy of the excited atomic state is larger than 2Φ.

There is no simple formalism describing the potential emission. Measurements can be found in literature such as shown in Fig. 7.17 for the energy distribution of the secondary electrons emitted during low-energy bombardment. At constant ion energy, both the yield and the mean energy of the electrons increase with decreasing ion mass, due to the increasing ionization energy. Estimating a total SEEC yield from Fig. 7.17 yields values between about $2.5 \cdot 10^{-2}$ and $2.5 \cdot 10^{-1}$. A very rough empirical relation for clean metal surfaces is

$$\gamma_e = 0.016 \cdot (E_i - 2\Phi)/eV. \tag{7.28}$$

Potential emission is characteristic at ion energies from floating plasmas. However, it is important to note that the potential emission of secondary electrons depends critically on the

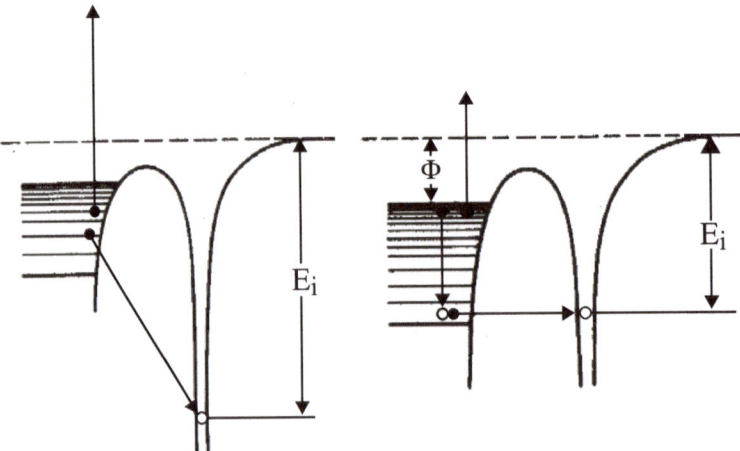

Figure 7.16: Two possible processes of potential emission of secondary electrons: Auger neutralization (left) and resonance neutralization (right).

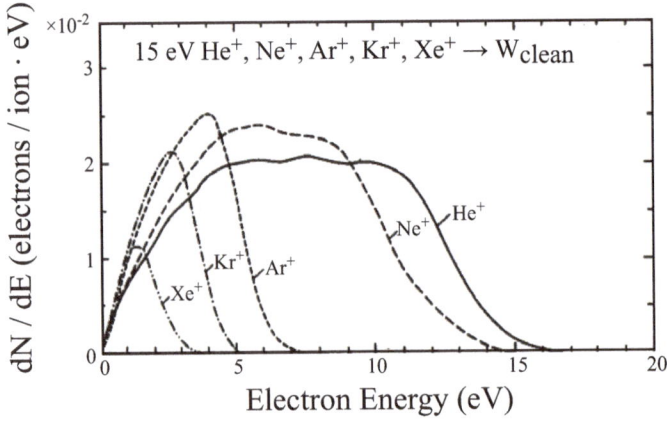

Figure 7.17: Energy distribution of secondary electrons due to potential emission from a clean tungsten surface, by different low energy ions.

surface state. Adsorbates on the surface, such as oxidation, may increase the SEEC significantly. As standard plasma processes do not make use of UHV technology and the process gases do normally not comply with extreme requirements of purity, the surface state during a plasma process is mostly unknown, as it results from a balance of different chemical reactions and ion induced processes as described above.

Kinetic emisson of secondary electrons is attributed to the creation of core vacancies in surface or bulk atoms due to electronic collisions. Transitions from higher levels cause cascades of Auger electrons. From the primary process, the SEEC is expected to be proportional to the electronic stopping cross section. According to Section 7.1 for sufficiently low energies, it this scales with the ion velocity

$$\gamma_e \sim S_i(E_0) \sim \sqrt{E_0}. \tag{7.29}$$

This is roughly confirmed in Fig. 7.18. In contrast to potential emission, the SEEC increases with the ion mass due to increased electronic stopping. Kinetic emission becomes dominant at energies well above 100 eV, so that it determines the SEEC in plasma-surface interaction only with biased substrates. It is seen from Fig. 7.18 that also kinetic emission is significantly influenced by surface impurities.

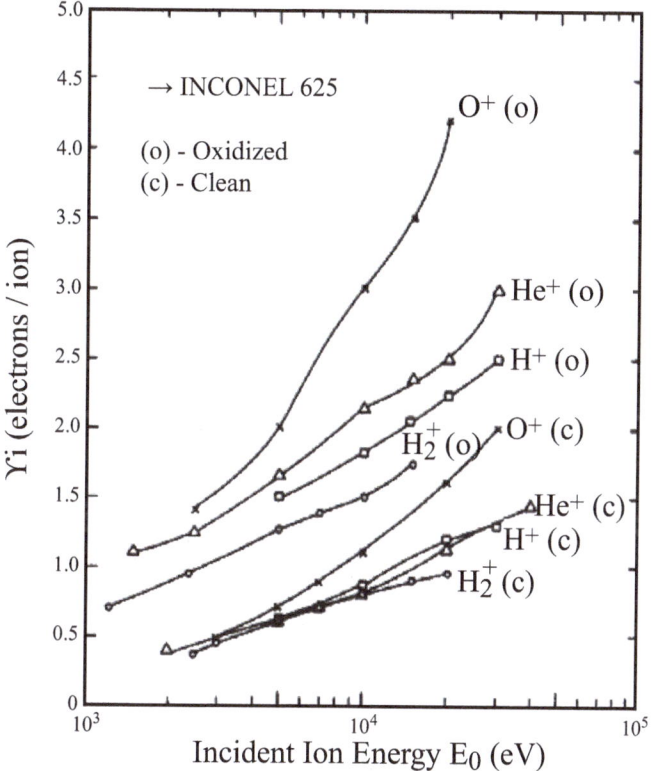

Figure 7.18: SEEC by kinetic emission for different ions incident on a clean and an oxidized surface of a nickel-based structural alloy.

CHAPTER 8

Particle Waves and Resonances

8.1   ELECTRON OSCILLATIONS

In order to get used to the formalism employed in the following, we will start with a more rigorous derivation of the electron plasma frequency. As shown in Fig. 1.3, the ions are assumed to be immobile. Then, in absence of a magnetic field, the electron momentum balance reads for an homogeneous plasma

$$m_e n \left[\frac{\partial \vec{u}_e}{\partial t} + (\vec{u} \cdot \vec{\nabla}) \vec{u} \right] = -en\vec{E} \tag{8.1}$$

and the equation of continuity

$$\frac{\partial}{\partial t} n + n\vec{\nabla} \cdot \vec{u} = 0. \tag{8.2}$$

Applying small time-dependent disturbances (index 1) to the velocity, $u = u_e + u_1$, the density, $n = n_e + n_1$, and the electric field $E = E_0 + E_1$, and treating only the time-dependent fraction, results in

$$m_e \left[\frac{\partial \vec{u}_1}{\partial t} + (\vec{u}_1 \cdot \vec{\nabla}) \vec{u}_1 \right] = -e\vec{E}_1. \tag{8.3}$$

The second term is second order in the disturbance and can be omitted. Eq. (8.2) becomes

$$\frac{\partial}{\partial t} n_1 + n_e \vec{\nabla} \cdot \vec{u}_1 + n_1 \vec{\nabla} \cdot \vec{u}_1 = 0 \tag{8.4}$$

as the plasma is homogeneous. Again, the third term is of second order. The Poisson equation reads

$$\vec{\nabla} \cdot \vec{E}_1 = -\frac{e}{\varepsilon_0} n_1. \tag{8.5}$$

Now we write the disturbances as periodic, i.e.,

$$n_1 = n_{10} e^{i(kx - \omega t)} \tag{8.6}$$

and correspondingly for u and E. Then, Eqs. (8.3)–(8.5) read

$$-i m_e \omega u_1 = -e E_1 \tag{8.7}$$

$$-i \omega n_1 = -n_e i k u_1 \tag{8.8}$$

$$\varepsilon_0 i k E_1 = -e n_1 \tag{8.9}$$

Elimination of u_1, E_1, and u_1 yields the electron plasma frequency of Eq. (1.12).

The described oscillation does not propagate, as regions next to the disturbed region feel no electric field due the local charge balance.

8.2   ELECTRON WAVES

For an inhomogeneous plasma, the pressure gradient has to be added to Eq. (8.1). Assuming that the oscillations are faster than the temperature equilibration, the adiabatic equation of state applies:

$$p = \frac{2+N}{N} nkT, \tag{8.10}$$

where N denotes the number of the degrees of freedom. For the oscillations in one direction as shown in Fig. 8.1, $N = 1$ and thereby $p = 3nkT$. Then, the momentum balance becomes

$$m_e n \left[\frac{\partial \vec{u}}{\partial t} + (\vec{u} \cdot \vec{\nabla})\vec{u} \right] = -en\vec{E} - 3kT_e\vec{\nabla}n. \tag{8.11}$$

As above,

$$-in_e m_e \omega u_1 = -en_e E_1 - 3kT_e ik n_1 \tag{8.12}$$

whereas Eqs. (8.8) and (8.9) remain unchanged. Elimination results in

$$\omega^2 = \omega_{pe}^2 + \frac{3kT_e}{m_e}k^2 = \omega_{pe}^2 + \langle v_{e,th}^2\rangle k^2. \tag{8.13}$$

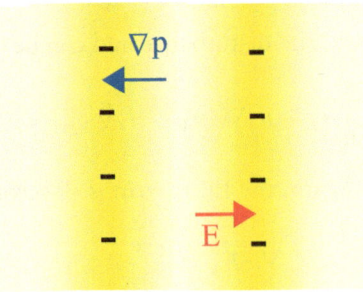

Figure 8.1: Electron waves. A pressure gradient induces an electric field.

This dispersion relation is shown in Fig. 8.2. The propagation velocity

$$v_g = \frac{d\omega}{dk} \tag{8.14}$$

is small at small k corresponding to the small gradients associated with the large wavelength, and tends toward

$$v_g|_{k\to\infty} = \sqrt{\langle v_{e,th}^2\rangle}. \tag{8.15}$$

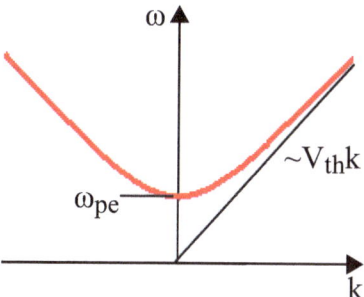

Figure 8.2: Dispersion of electron waves. for short wavelength, they are driven thermally.

Thus, the propagation at sufficiently small wavelength is given by the thermal velocity of the electrons, the electron waves are thermally driven.

8.3   ION WAVES

Similar electrostatic waves can propagate in the ion subsystem. The momentum balance reads

$$m_i n \left[\frac{\partial \vec{u}}{\partial t} + (\vec{u} \cdot \vec{\nabla}) \vec{u} \right] = -e n \vec{\nabla} \Phi - 3k T_i \vec{\nabla} n \tag{8.16}$$

with the electrostatic potential Φ, and the Fourier transform

$$- i n_e m_i \omega u_1 = -e n_e i k \Phi_1 - 3k T_i i k n_1. \tag{8.17}$$

However, now the highly mobile electrons are able to follow the oscillations of the ions. Using the Boltzmann relation with n_- denoting the electron density under the influence of the potential by the ions,

$$n_- = n_e + n_{1e} = n_e \exp\left(\frac{e\Phi_1}{kT_e} \right) = n_e \left(1 + \frac{e\Phi_1}{kT_e} + \ldots \right). \tag{8.18}$$

For an approximate solution, we assume neutrality also in the fluctuations, which holds for wavelength being small compared to the Debye length. Then

$$n_{1e} = n_e \frac{e\Phi_1}{kT_e} = n_1 \tag{8.19}$$

while Eq. (8.8) still holds, now for the ions. Elimination of n_1, υ_1, Φ_1 from Eqs. (8.8), (8.17) and (8.19) yields

$$\omega^2|_{k\to 0} = k^2 \left(\frac{kT_e}{m_i} + \frac{3kT_i}{m_i} \right). \tag{8.20}$$

Furthermore, the ion wave is driven by the temperature. In cold plasmas, the propagation is only given by the electron temperature. Cold electrons would entirely follow the ions and thus maintain neutrality everywhere, so that the wave cannot propagate. A rigorous treatment yields for large wave vectors the ion plasma frequency

$$\omega^2|_{k\to0} = \omega_{pi}^2 = \frac{n_e e^2}{\varepsilon_0 m_i} \tag{8.21}$$

so that the oscillation does again not propagate.

8.4   ELECTRON OSCILLATIONS IN MAGNETIC FIELDS

As for the transport phenomena in the previous chapters, the formation of particle waves parallel to the magnetic field is not influenced by the field. New phenomena, however, arise for a transport perpendicular to the magnetic field. The momentum balance for the fluctuations reads, now for a homogeneous plasma

$$m_e \frac{\partial \vec{u}_1}{\partial t} = -e(\vec{E}_1 + \vec{u}_1 \times \vec{B}). \tag{8.22}$$

With the geometry of Fig. 8.3, the Fourier transform equation reads, as $E \| k$

$$-i m_e \omega u_{1x} = -e E_1 - e u_{1y} B \tag{8.23}$$

$$-i m_e \omega u_{1y} = e u_{1x} B \tag{8.24}$$

$$-i m_e \omega u_{1z} = 0. \tag{8.25}$$

Inserting Eq. (8.24) into Eq. (8.23),

$$i m_e \omega u_{1x} = e E_1 + i \frac{e^2 B^2}{m_2 \omega} u_{1x}. \tag{8.26}$$

From the particle balance,

$$\omega n_1 = k n_e u_{1x}. \tag{8.27}$$

The Poisson equation is written as

$$\varepsilon_0 i k E_1 = -e n_1. \tag{8.28}$$

Combining Eqs. (8.26)–(8.28) yields

$$\left(1 - \frac{\omega_C^2}{\omega^2}\right) = \frac{\omega_{pe}^2}{\omega^2} \tag{8.29}$$

or

$$\omega^2 = \omega_{uh}^2 = \omega_{pe}^2 + \omega_C^2. \tag{8.30}$$

This is the so-called "upper hybrid" resonance frequency. There is no propagation of the wave, but in the resonance condition the oscillation of the electrons is amplified by the superimposed gyration. This is shown schematically in Fig. 8.4.

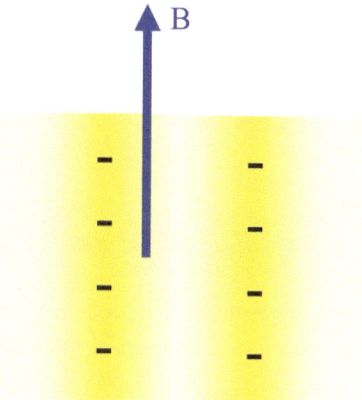

Figure 8.3: Electron oscillations perpendicular to a magnetic field.

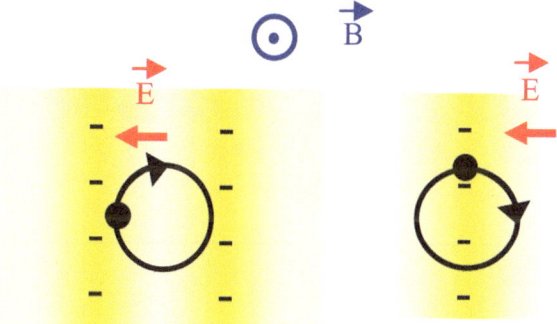

Figure 8.4: Upper hybrid resonance: the electron gyrates in phase with the electron oscillation. From left to right, the phase is advanced by 90°. In both cases, the electron is accelerated.

8.5 ION WAVES IN MAGNETIC FIELDS

Ion waves (Section 8.2) are characterized by the fact that the highly mobile electrons follow the ion oscillations. In a magnetic field which is perpendicular to the direction of propagation, the ions with their large gyration radii are little influenced by the field, whereas the transverse motion of the electrons is suppressed. However, if the angle between the magnetic field and the wave vector is denoted by Θ already a small deviation from $\Theta = 90°$ will allow the electrons again to adjust to the ion system.

The momentum balance reads now

$$m_i \frac{\partial \vec{u}_1}{\partial t} = -e\vec{\nabla}\Phi_1 + e\vec{u}_1 \times \vec{B}. \tag{8.31}$$

For $\Theta \neq 90°$, the electron kinetics can be treated as in Section 8.2, and the dispersion relation can be derived in analogy to Section 8.3. The result becomes

$$\omega^2 = \omega_{Ci}^2 + k^2 \frac{kT_e}{m_i} \tag{8.32}$$

with the ion cyclotron frequency

$$\omega_{Ci} = \frac{eB}{m_i}. \tag{8.33}$$

Consequently, these propagating waves are called "ion cyclotron waves." The gyration of the ions can be understood to be accelerated by ambipolar diffusion, which is driven by the electrons.

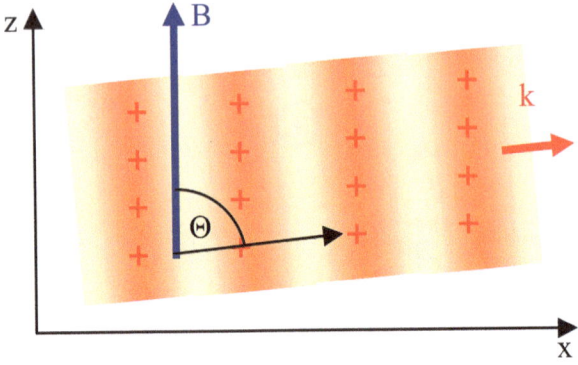

Figure 8.5: Ion waves in a magnetic field with a propagation nearly perpendicular to the field.

In case of $\Theta = 90°$, the set of the momentum balance, particle balance and Poisson equations has to be solved both for electrons and ions. The electrons are pinned to the magnetic field lines. Consequently, propagation of a wave is not possible. A resonance condition is fulfilled when the ions oscillate in phase with the gyration of the electrons, i.e., (see Fig. 8.6)

$$u_{1x,i} = u_{1x,e}. \tag{8.34}$$

With this condition, the resonance frequency results as

$$\omega = \omega_{lh} = \sqrt{\omega_{Ce} \cdot \omega_{Ci}} \tag{8.35}$$

which is called the "lower hybrid" resonance frequency.

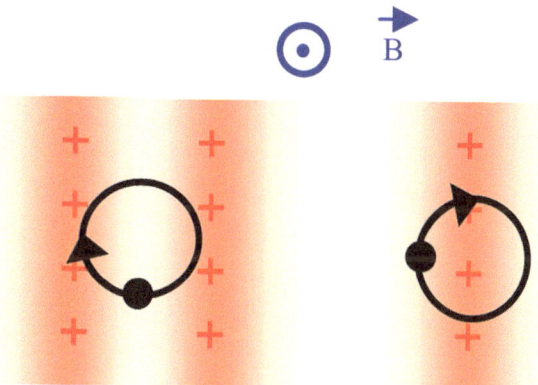

Figure 8.6: Lower hybrid resonance: the electron gyrates in phase with the ion oscillation. From left to right, the phase is advanced by 90°. In both cases, the electron is accelerated.

CHAPTER 9

Electromagnetic Waves

Starting point are the Maxwell equations

$$\vec{\nabla} \times \vec{E} = -\frac{\partial}{\partial t} \vec{B} \tag{9.1}$$

$$\vec{\nabla} \times \vec{B} = \mu_0 \vec{j} + \varepsilon_0 \mu_0 \frac{\partial}{\partial t} \vec{E}, \tag{9.2}$$

where j denotes the electrical current density here. Combining yields

$$-\vec{\nabla} \times (\vec{\nabla} \times \vec{E}) = \mu_0 \frac{\partial}{\partial t} \vec{j} + \varepsilon_0 \mu_0 \frac{\partial^2}{\partial t^2} \vec{E} \tag{9.3}$$

or

$$-\vec{\nabla}(\vec{\nabla} \cdot \vec{E}) + \vec{\nabla}^2 \vec{E} = \mu_0 \frac{\partial}{\partial t} \vec{j} + \frac{1}{c^2} \frac{\partial^2}{\partial t^2} \vec{E}. \tag{9.4}$$

Writing

$$\vec{E} = \vec{E}_0 e^{(\vec{k}\vec{x} - \omega t)} \tag{9.5}$$

$$\vec{j} = \vec{j}_0 e^{i(\vec{k}\vec{x} - \omega t)} \tag{9.6}$$

results in the wave equation

$$\vec{k}(\vec{k} \cdot \vec{E}) - k^2 \vec{E} = -i\omega \mu_0 \vec{j} - \frac{\omega^2}{c^2} \vec{E}. \tag{9.7}$$

9.1 NON-MAGNETIZED PLASMA

For transversal EM waves, as $k \perp E$

$$k^2 \vec{E} = i\omega \mu_0 \vec{j} + \frac{\omega^2}{c^2} \vec{E}. \tag{9.8}$$

For the response of the electrons to the electric field in a homogeneous plasma, we write the electron current

$$\vec{j} = -n_e e \vec{v}_e \tag{9.9}$$

and the momentum balance by

$$n_e m_e \frac{\partial}{\partial t} \vec{v}_e = -n_e e \vec{E} - n_e m_e v_{ce} \vec{v}_e, \tag{9.10}$$

where the second-order term on the left-hand side has again been neglected, and the collision tern is introduced. As above,

$$- m_{ei}\omega\vec{v}_e = -e\vec{E} - m_e v_{ce}\vec{v}_e. \tag{9.11}$$

All above vectors point into the same direction, so we turn to scalars

$$v_e = \frac{e}{m_e}\frac{E}{i\omega + v_{ce}}. \tag{9.12}$$

Inserting this expression into Eq. (9.9) and then into Eq. (9.8), one gets

$$k^2 = -i\omega\mu_0\frac{n_e e^2}{m_e}\frac{1}{i\omega + v_{ce}} + \frac{\omega^2}{c^2} \tag{9.13}$$

or

$$k^2 c^2 = \omega^2\left(1 - \frac{\omega_{pe}^2}{\omega^2}\frac{\omega}{\omega - i v_{ce}}\right). \tag{9.14}$$

Figure 9.1 shows the dispersion relation with the absorption neglected. The result demonstrates that in the limit of high frequencies, the vacuum dispersion relation is obtained, as the electrons can no longer follow. Toward low freqency, propagation is retarded, until finally only the non-propagating electron wave with the plasma freqency, which is slightly reduced due to collisions, can exist.

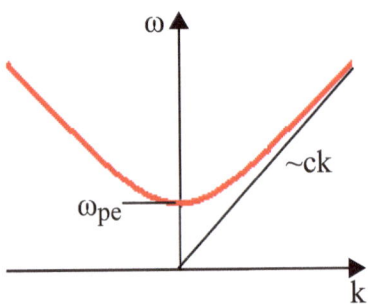

Figure 9.1: Dispersion of electromagnetic waves in a non-magnetized plasma without collisions. For short wavelength, the wave propagates like in vacuum.

For the electromagnetic wave, the group or propagation velocity is given by Eq. (8.14). The phase velocity is

$$v_{ph} = \frac{\omega}{k}. \tag{9.15}$$

Equation (9.15) shows that the phase velocity is slightly reduced due to collisions. The refractive index is defined as

$$n = \frac{c}{v_{ph}} = \frac{ck}{\omega}. \tag{9.16}$$

Thus, in Eq. (9.14),

$$n^2 = 1 - \frac{\omega_{pe}^2}{\omega^2} \frac{\omega}{\omega - i v_{ce}}. \tag{9.17}$$

The imaginary part of n determines adsorption. Thus, in the present the absorption is caused only by collisions.

A so-called "cut-off" occurs for

$$k \to 0 \qquad \text{or} \qquad n \to 0 \tag{9.18}$$

which corresponds to an infinite wavelength so that a wave cannot propagate. If collisions are neglected, the cutoff frequency of the above wave is given by

$$\omega_{co} = \omega_{pe}. \tag{9.19}$$

At a frequency lower than the plasma frequency, n becomes purely imaginary (in the absence of collisions) so that the wave is absorbed. Writing

$$n = \alpha + i\beta \tag{9.20}$$

with the (real) refractive index α and the absorption coefficient β, the amplitude of the wave decays exponentially according to

$$E = E_0 e^{i\omega\left(\frac{nx}{c} - t\right)} = E_0 e^{i\omega\left(\frac{\alpha x}{c} + i\frac{\beta x}{c} - t\right)} = E_0 e^{-\omega\frac{\beta x}{c}} e^{i\omega\left(\frac{\alpha x}{c} - t\right)} \tag{9.21}$$

with a characteristic penetration depth ("skin depth")

$$x_{sk} = \frac{c}{\beta \omega} \tag{9.22}$$

This is displayed schematically in Fig. 9.2. In the present case and again without collisions, for $\omega < \omega_{pe}$ according to Eq. (9.17)

$$x_{sk} = \frac{c}{\sqrt{\omega_{pe}^2 - \omega^2}} \xrightarrow{\omega \to 0} \frac{c}{\omega_{pe}}. \tag{9.23}$$

which illustrates that the penetration depth into the plasma is independent of the frequency of the wave at sufficiently low frequencies.

The cut-off condition can be rewritten in terms of a maximum electron density above which a wave with a given frequency is unable to penetrate.

$$n_{e,\max} = \frac{\varepsilon_0 m_e}{e^2} \omega^2. \tag{9.24}$$

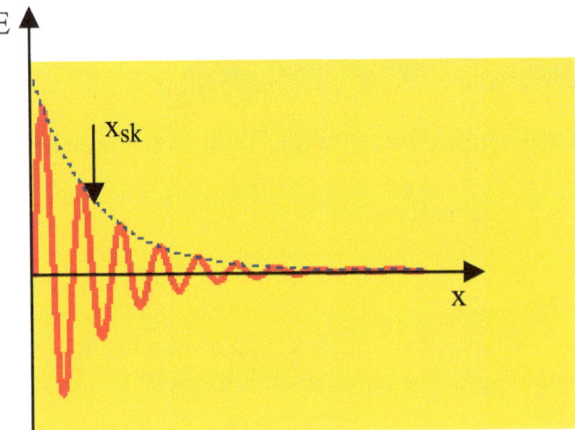

Figure 9.2: Evanescent wave in a plasma at a frequency below the cutoff frequency. x_{sk} denotes the skin depth.

For example, microwave heating of a plasma with the standard frequency of $f = 2.45$ GHz results in a maximum density of $n_e \approx 6 \cdot 10^{10}$ cm^{-3}. Due to the quadratic dependence on the frequency in Eq. (9.24), it is desirable to use very high frequencies if high plasma densities shall be achieved.

Resonances occur for

$$k \to \infty \quad \text{or} \quad n \to \infty. \tag{9.25}$$

Formally, according to Eq. (9.17), this would be fulfilled by $\omega \to 0$. However, this solution does not exist according to the dispersion relation, as the wave becomes evanescent for small ω.

9.2   MAGNETIZED PLASMA

In treating electromagnetic wave propagation in the presence of a magnetic field, we start with the geometry of Fig. 9.3, with $k \perp B$ and $E \| B$. Then, the electrons which are accelerated by the wave are not influenced by the magnetic field, and the results of the previous section apply. The corresponding wave, which is also denoted as P wave (as $E \| B$), is called the "ordinary wave."

With $k \perp B$, but $E \perp B$ (see Fig. 9.4) and neglecting collisions, the momentum balance is

$$m_e \frac{\partial}{\partial t} \vec{v}_e = -e\vec{E} - e(\vec{v}_e \times \vec{B}). \tag{9.26}$$

Evaluation from this v_x and v_y, and inserting into the wave Equation (9.7) yields

$$(\omega^2 - k^2 c^2)\vec{E} + c^2 \vec{k}(\vec{k} \cdot \vec{E}) = -\frac{i\omega}{\varepsilon_0}\vec{j} = \frac{i n_e e \omega}{\varepsilon_0}\vec{v}_e. \tag{9.27}$$

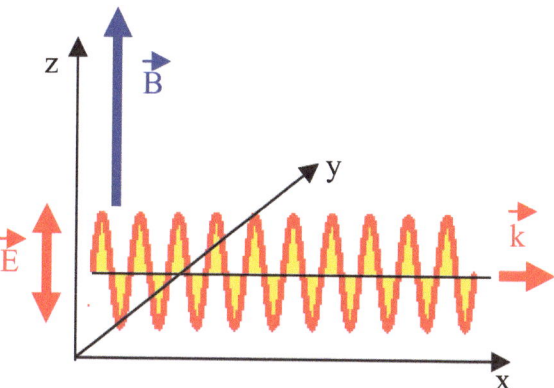

Figure 9.3: Electromagnetic "*P*" wave propagating perpendicular to the magnetic field, with the electric field vector parallel to the magnetic field.

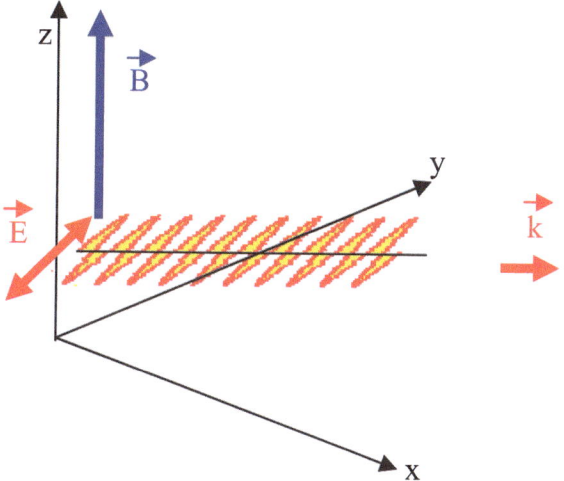

Figure 9.4: Electromagnetic wave propagating perpendicular to the magnetic field, with the electric field vector perpendicular to the magnetic field.

Again the velocity components are inserted on the right-hand side. Lengthy rearrangements then yield the dispersion relation for the "extraordinary" wave

$$k^2 c^2 = \omega^2 \left(1 - \frac{\omega_{pe}^2}{\omega^2} \frac{\omega^2 - \omega_{pe}^2}{\omega^2 - \omega_{uh}^2} \right) \tag{9.28}$$

with the upper hybrid frequency of Eq. (8.30). With Eq. (9.16), the inverse square of the refractive index is plotted vs. the EM wave frequency in Fig. 9.5.

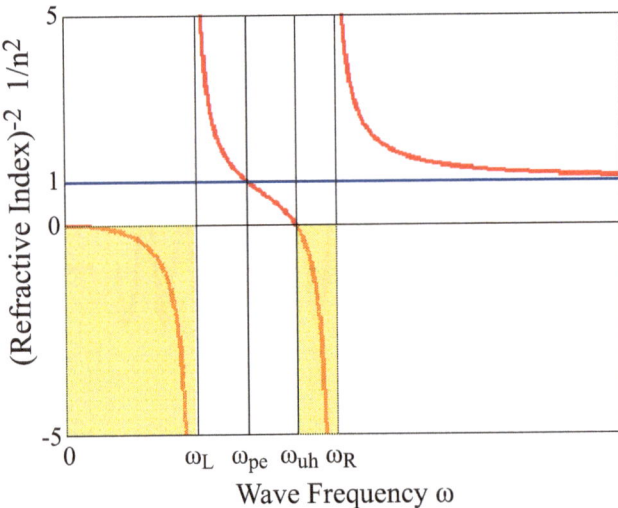

Figure 9.5: Inverse square of the refractive index of the wave defined in Fig. 9.4 for a cyclotron frequency of 3/4 of the plasma frequency. A resonance occurs at the upper hybride frequency, cutoffs occur at two frequencies (indexed by L and R) due to left- and right-hand gyration of the electrons. The regimes of wave evanescence are shaded.

The cutoff condition of Eq. (9.18) results after replacing the upper hybrid frequency by the plasma and cyclotron frequencies, as follows:

$$\frac{\omega_{pe}^2}{\omega^2}\frac{1}{1-\omega_{Ce}^2\frac{1}{\omega^2-\omega_{pe}^2}} = 1. \tag{9.29}$$

From this,

$$1 - \frac{\omega_{pe}^2}{\omega^2} = \pm\frac{\omega_{Ce}^2}{\omega} \tag{9.30}$$

and

$$\omega_{R,L} = \omega_{pe}\left(\sqrt{1+\frac{\omega_{Ce}^2}{4\omega_{pe}^2}} \pm \frac{\omega_{Ce}}{2\omega_{pe}}\right). \tag{9.31}$$

The $+$ and $-$ signs correspond to the left-hand (L) and right-hand (R) cutoffs. This is explained in the following way: the linearly polarized wave with the electric field vector perpendicular to the magnetic field can be decomposed into two equivalent waves of circular polarization, denoted as L and R waves (see Fig. 9.6). The R circular wave acts on the gyrating electrons in the opposite way then the L wave. From Eq. (9.30), the two cutoff frequencies differ by the cyclotron frequency.

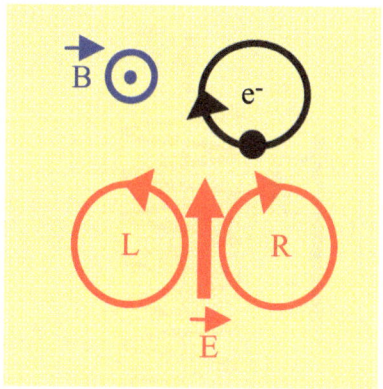

Figure 9.6: A linearly polarized wave is equivalent to two circularly polarized waves L and R. For $E \perp B$, only one of them (R) can interact resonantly with the gyrating electron.

A resonance occurs at

$$\omega = \omega_{uh} \tag{9.32}$$

which means that the wave couples to the electron oscillations in a magnetic field as described in Section 8.3.

If the wave propagates parallel to the magnetic field (see Fig. 9.7), the results derived in analogy to the above is

$$k^2 c^2 = \omega^2 \left(1 - \frac{\omega_{pe}^2}{\omega^2} \frac{\omega}{\omega \mp \omega_{ce}} \right) \tag{9.33}$$

which indicates a specific solution for each of the two circularly polarized partial waves into which the linearly polarized wave can be decomposed. The refractive index plot is shown in Fig. 9.8.

The cutoff condition resulting from Eq. (9.33) are identical to those given in Eq. (9.31).

A resonance occurs only for the R wave, for which of the E vector rotates in the same direction as the gyrating electrons. This resonance limits the regime of the so-called "whistler mode" propagation. For the whistler mode, there is an analogon in acoustics. The curvature of the refractive index vs. the frequency (and thereby the curvature of the dispersion relation) means that the propagation velocity, Eq. (8.14), is smaller for lower frequencies.

The propagation of electromagnetic waves is often illustrated in a so-called CMA diagram (after Clemmow, Mullaly, and Allis). Figure 9.9 displays such a diagram for the above cases with coplanar k, E, and B vectors, i.e., for the geometries given in Figs. 9.3 and 9.7, with the cutoff results of Eqs. (9.19) and (9.31). The coordinates of the CMA diagram are conveniently proportional to the electron density of the plasma and the magnetic field.

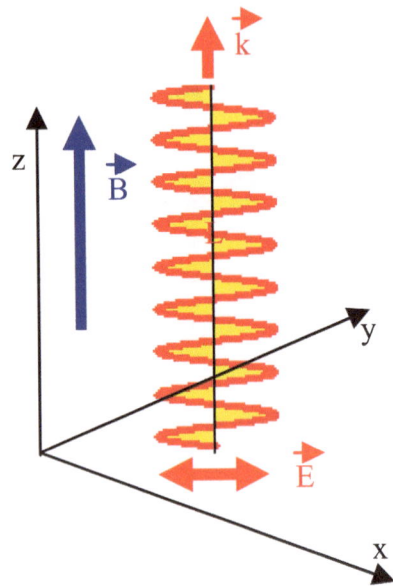

Figure 9.7: Electromagnetic wave propagating parallel to the magnetic field.

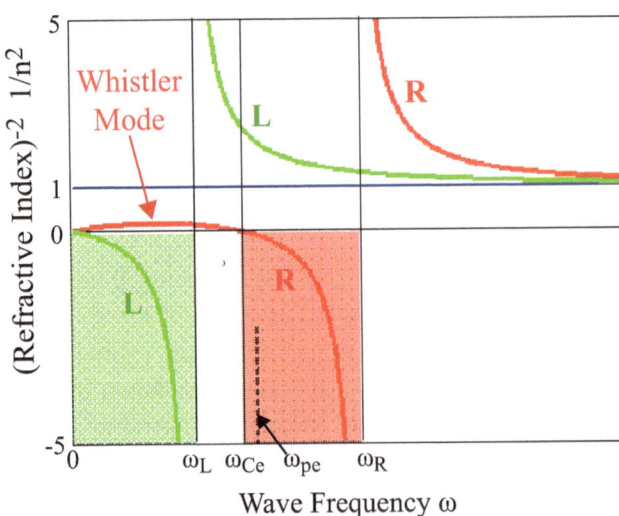

Figure 9.8: Inverse square of the refractive index of the wave defined in Fig. 9.6 for a cyclotron frequency of 3/4 of the plasma frequency. For the R and L polarized waves, cutoffs occur at different frequencies. Only the R wave shows a resonance at the cyclotron frequency. The regimes of wave evanescence are shaded.

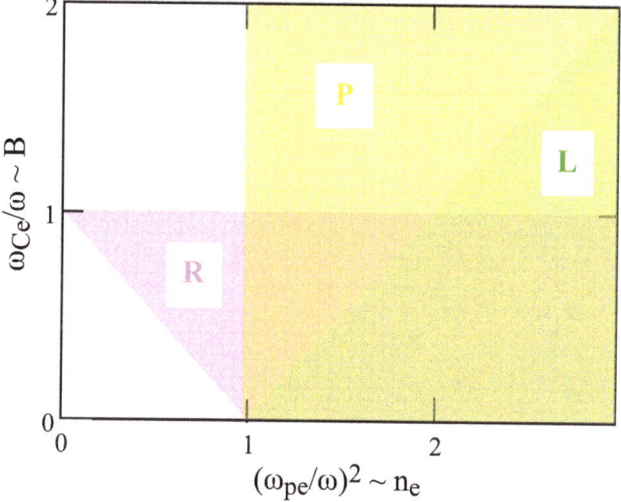

Figure 9.9: Propagation regimes (colored) of the P, R, and L waves for $k \times [E \times B] = 0$, in a CMA diagram. The border lines are given by the cutoff relations, Eqs. (9.19) and Eq. (9.29).

CHAPTER 10

Plasma Modeling

10.1 GLOBAL MODEL

The following section will describe a very simple "global" plasma model, which is of zero dimensionality and results from particle and power balances only. We restrict ourselves to a plasma with only one species of neutral gas density n_0 and ions, and a low degree of ionization

$$\alpha_i = \frac{n_e}{n_0}. \tag{10.1}$$

According to Fig. 1.1, a plasma is assumed in which the ions are homogeneously generated by electron impact. In the simplest case, only ionization and wall losses of the charged particles are taken into account. Ionization is assumed to be only due to the thermalized electrons of the plasma (electron "beams" are neglected). For the losses to the wall, a mean ion confinement time τ_i is defined by the loss of charged particles per unit volume and unit time. Then, the ion balance equation reads

$$\frac{\partial n_i}{\partial t} = n_e n_0 \langle \sigma \upsilon \rangle_i - \frac{n_i}{\tau_i}, \tag{10.2}$$

where $< \sigma \upsilon >_i$ is the ionization rate coefficient according to Eq. (2.1), averaged over the electron velocity distribution. If ion generation would be switched off, the ion density would decay exponentially according to $n_i \sim \exp(-t/\tau_i)$.

In stationary state and due to quasi-neutrality, Eq. (10.2) simplifies to

$$\tau_i n_0 \langle \sigma \upsilon \rangle_i = 1. \tag{10.3}$$

From this equation, the ion density has dropped out, demonstrating that the ion density is not determined by the ion balance. In low-pressure plasmas with an electron temperature of a few eV, only the high-energy Maxwellian tail contributes to ionization, if the ionization potential is significantly higher than the electron temperature. Then, the rate coefficient is a steeply increasing function of the electron temperature, as seen in Fig. 2.9.

For the loss of ions, the area integrated flux onto the wall has to be equal to the volume-integrated loss of density. With Eq. (6.21),

$$j_i A = e^{-1/2} n_i \upsilon_B = \frac{n_i}{\tau_i} V. \tag{10.4}$$

With the characteristic dimension of the vacuum chamber (Eq. (5.20)) the ion confinement time becomes

$$\tau_i = de^{1/2}\sqrt{\frac{m_i}{kT_e}} \tag{10.5}$$

and thus from Eq. (10.3)

$$e^{1/2}dn_0\langle\sigma\upsilon(kT_e)\rangle_i = \sqrt{\frac{kT_e}{m_i}} \tag{10.6}$$

where the left-hand side varies much steeper with the electron temperature than the right-hand side. Consequently, the plasma adjusts the electron temperature so that Eq. (10.6) is fulfilled. At given ion species, the electron temperature decreases at increasing filling gas pressure, but the dependence is rather weak. At fixed pressure, it is determined by the geometry of the chamber.

Whereas the ion balance thus determines the electron temperature, the ion density is determined by the power balance. Without specifying any source of the electrical power, P, which is fed into the discharge (see Fig. 10.1), a fraction P_p of it is consumed by the bulk processes of ionization, dissociation, excitation ... Here, we assume that ionization is the main consumer and neglect the other processes. Then, the absorbed power is given by

$$P_p = n_e n_0 V \langle\sigma\upsilon(kT_e)\rangle_i U_i. \tag{10.7}$$

If P_p is assumed to be constant, the plasma density results directly with the electron temperature defined by Eq. (10.6). The plasma density is proportional to the absorbed power, inversely proportional to the gas pressure and the ionization energy, and strongly decreasing at increasing electron temperature.

Refining this simplest global model, we consider that for a floating plasma, the ion acceleration in the sheath has also to be provided by the power which is fed into the plasma. The power consumed for ion acceleration is

$$P_i = j_i E_i A \tag{10.8}$$

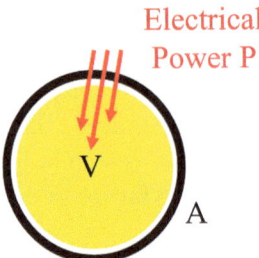

Figure 10.1: Electrical power P fed into the container to sustain the plasma.

with the ion flux and the ion energy given by Eqs. (6.21) and (6.26), respectively, so that the total input power results as

$$P = P_i + P_p. \tag{10.9}$$

With Eqs. (10.7) and (10.8), the plasma density is calculated according to

$$n_e = P\left(n_0 V \langle \sigma \upsilon (kT_e) \rangle_i U_i + \frac{A(kT_e)^{3/2}}{2\exp(1/2)m_i^{1/2}}\left(1 + \log\frac{m_i}{2\pi m_e}\right)\right)^{-1}. \tag{10.10}$$

Figure 10.2 shows a set of results for an argon plasma. At the given conditions, the maximum degree of ionization is approx. 10^{-3}. The electron density increases almost linearly with the pressure, whereas the electron temperature drops. The electron temperature is $2\ldots5$ eV at pressures above 0.1 Pa, which are typical for low-pressure plasma processing. Correspondingly, the ion energies are between 10 and 30 eV. The ion current varies by less than a factor of two for pressures between 0.1 and 10 Pa. The input power is mainly consumed for ion acceleration except for the highest pressures.

The results become unrealistic at the lowest pressures, as it will hardly be possible to fed the given input power into the gas at such low density, using standard means of plasma generation (see Chapter 11).

At the higher pressures, the results become increasingly questionable as collisions become important. At 1 Pa, the mean free path length is about 2 cm and clearly lower than the connection length. In this context, it is interesting to compare the ion flux which would result from ambipolar diffusion to the Bohm flux which has been employed for the global model. From Eq. (5.31), the ambipolar diffusion flux is with the mean free path length λ_c and the mean ion velocity $\upsilon_{i,th}$

$$j_a \approx D_a \frac{n_e}{d} = \frac{kT_e}{m_i} \frac{\lambda_c}{d} \frac{n_e}{\upsilon_{i,th}} = \sqrt{\frac{kT_e}{m_i}} \sqrt{\frac{kT_e}{kT_i}} \frac{1}{\sqrt{3}} \frac{\lambda_c}{d} n_e \approx j_i \sqrt{\frac{kT_e}{kT_i}} \frac{\lambda_c}{d}. \tag{10.11}$$

Thus, the ambipolar diffusion becomes rate-controlling if

$$\sqrt{\frac{kT_e}{kT_i}} \frac{\lambda_c}{d} < 1. \tag{10.12}$$

For the above example, the ratio of the electron and ion temperatures varies only slowly with pressure and is about 50. Thus, the ambipolar diffusion flux is larger than the Bohm flux as long as the mean free path length is larger than about $d/7$, which corresponds to a pressure of about 3 Pa. Thus, above this pressure the results become questionable as collisions have been neglected. However, a similar global model could also be formulated for the collisional case, with the ambipolar diffusion flux, Eq. (5.28), replacing the Bohm flux in Eq. (10.4).

A general remark of care has to be added, which relates not only to global modeling. In low-temperature plasmas, the electron temperature is mostly significantly below the ionization

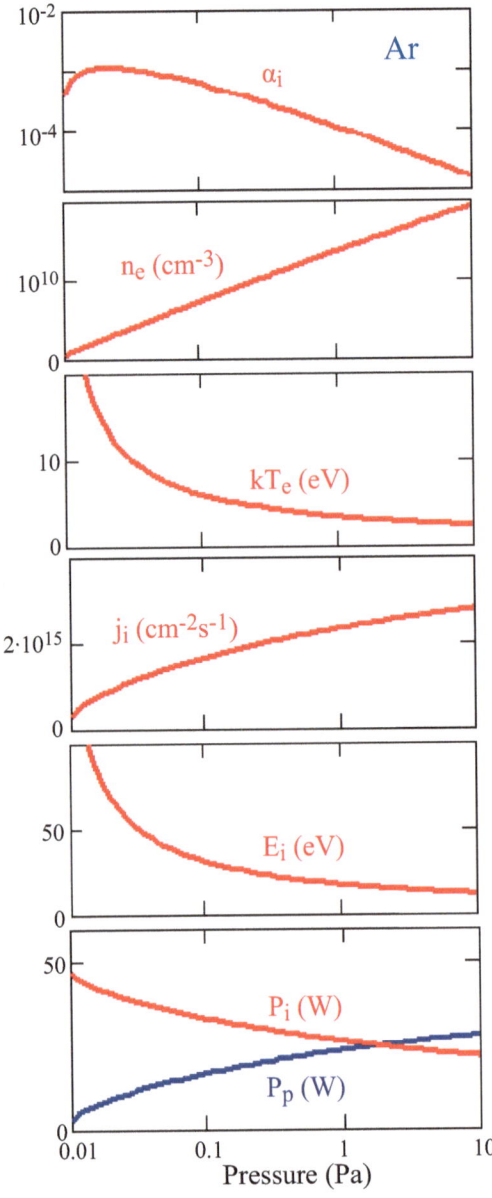

Figure 10.2: Degree of ionization, electron density and temperature, ion flux and energy, and powers consumed for ionization (blue) and ion acceleration in the sheath (red), in an argon plasma at dfferent gas pressures, as obtained from the stationary global model. The total input power P is 50 W, the container volume 20 l and the surface area 0.4 m^2, resulting in a characteristic connection length of 5 cm. Note that above 1 Pa, the plasma becomes partly collisional.

potential, so that the high-energy tail of the Maxwellian energy distribution dominates (see Section 2.8). Then, the validity of modeling depends critically on the presence of a Maxwell distribution. In low-pressure plasmas, the number of electron-electron collisions may not be sufficient to establish a real Maxwellian distribution, in particular in the high-energy tail.

For the above, a constant gas density n0 has been assumed. In reality, part of the filling gas is "consumed" by the electron-induced processes. Further, the standard technical or experimental situation is a given gas influx into the device and a given pumping speed at the outlet, from which the gas pressure results as a balance. In stationary state, the balance of the feed gas density is then given by

$$\frac{j_0}{V} - \frac{S}{V}n_0 - n_e n_0 \sum_k \langle \sigma v \rangle_k = 0, \tag{10.13}$$

where j_0 and S denote the number of inflowing gas molecules per unit of time and the pumping speed at the outlet, respectively. (It should be noted that the gas influx is normally given in standard cubic centimetres per minute (sccm), which corresponds to $4.2 \cdot 10^{17} s^{-1}$). Further in Eq. (10.13), the sum extends over all processes of ionization, dissociation, etc., which consume the feed gas. Then, n_0 represents another variable in addition to n_e and kT_e, which can be obtained from solving the non-linear system of, e.g., Eqs. (10.6), (10.7), and (10.13).

10.2 REACTIVE PLASMAS

For so-called "reactive" plasmas, dissociation of molecules as well as chemical reactions in the plasma become important, so that the plasma, in addition to the feed gas and the electrons, will consist of ions and neutrals of different molecular and radical species. Then, the general balance equation of any heavy ion or neutral species k may be written as

$$\frac{\partial n_k}{\partial t} = n_e \sum_j n_j \langle \sigma v \rangle_{el,j \to k} - n_e \sum_j n_j \langle \sigma v \rangle_{el,k \to j}$$
$$+ \sum_{j,l} n_j n_l K_{chem,j+l \to k} - n_k \sum_{j,l} n_j K_{chem,k+j \to l} - \frac{n_k}{\tau_k}. \tag{10.14}$$

The first two terms of Eq. (10.14) denote the production and loss of species k due to electron-induced processes such as ionization, dissociation and electron attachment with the summations extending over all species from which k originates or which originate from k, respectively. Correspondingly, chemical reactions are taken into account by terms 3 and 4, which produce and deplete species k, respectively. As the chemical reactions occur between heavy particles, which are cold in low-temperature plasmas, they are governed by thermal chemical rate constants K_{chem}. Often, the rate constants, in particular between radicals of ionic or neutral state, are unknown so that estimates have to be inserted. Equations of type (10.14) can be added to the above global model equations for stationary or time-dependent solutions.

The number of relevant species even in relatively simple reactive plasmas can be very large. Figure 10.3 shows the result of global reactive modeling for the most dominant species in a methane plasma. The calculation has been performed in the so-called "plug-flow" approximation, i.e., considering a slab of volume which moves through the active volume at constant speed according to the flow rate of the feed gas. For the model, in total 42 electron-impact, neutral-neutral, ion-neutral, neutral-wall and ion-wall reactions have been taken into account. Other than in the simple global model, the electron kinetics in the applied RF field and the electron-induced ionization and dissociation of CH_4 have been treated by a Monte Carlo simulation

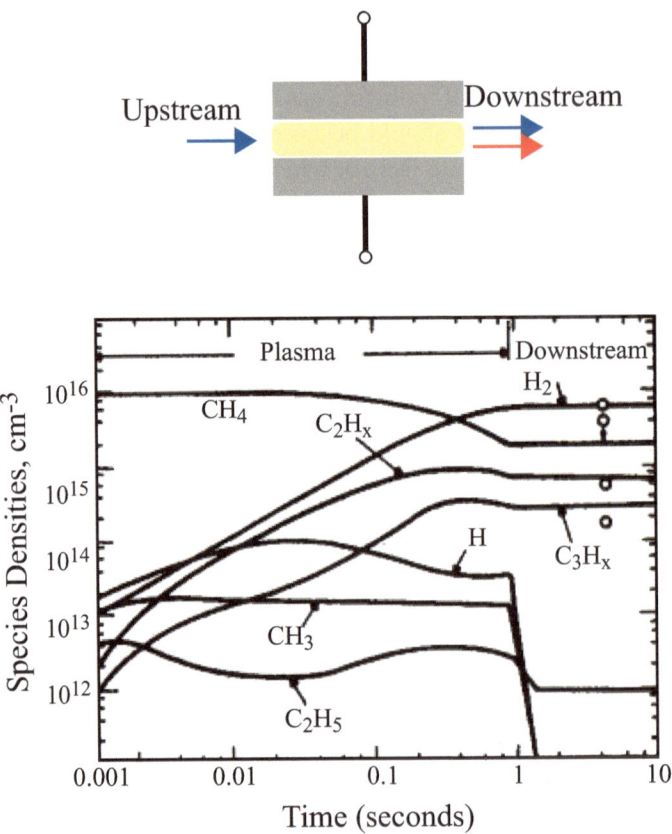

Figure 10.3: Evolution of the most dominant molecular species in a methane plasma, according to a time-dependent model calculation. A 2 MHz RF plasma is formed between two electrodes (see schematic), with the gas being fed from one side and pumped at the other side. The open points denote experimental results obtained by downstream mass spectroscopy. The gas pressure was 0.4 mbar, the electrode distance 2 cm, the plasma volume 125 cm^3, and the RF power density 120 mW/cm^3. (From L. E. Kline et al. [11].)

tracing a sufficiently large number of electron trajectories. The time-dependent evolution of species density in the moving slab corresponds to a spatial variation in the direction of the flow. In the present example of a comparatively low gas flow, the feed gas is largely depleted, so that the main gas component at the exit becomes molecular hydrogen, which is formed by abstraction and recombination reactions. From a methane plasma under the present conditions, a hydrocarbon film is deposited at the electrodes, which is assumed to be exclusively due to neutral and ionic CH_3 radicals which assumed wall sticking coefficients of 0.03 and 1, respectively. Then, reasonable agreement is obtained with mass spectrometric measurements of the main neutral species at the outlet.

The poor knowledge on surface processes adds considerably uncertainty to any kind of plasma modeling and computer simulation to be described below. A standard procedure is to insert "educated guesses" for effective sticking coefficients of the individual species for the boundary conditions. However, as pointed out in Section 7.6, the effective sticking of a particular species may depend on the wall fluxes of other species and thus not be constant. Then, as a refinement, wall reaction equations (similar to (7.23)) might be included into the model, with the wall coverages of reactive species as additional variables. Nevertheless, the processes which determine these equations are mostly unknown and have to be inferred in a simplified way.

10.3 FLUID MODELING

For fluid modeling of a plasma, the balance equations of the fluid model (Chapter 4) can be solved numerically. In the literature, sets of equations of widely different complexity are employed. Here, we restrict ourselves to a simple one-dimensional two-fluid description applied to the dynamics of a high-voltage plasma sheath (see Section 6.4) with a predefined plasma density and electron temperature. The equation of continuity (Eq. (4.16)) reads for the ions.

$$\frac{\partial n_i}{\partial t} + \frac{\partial}{\partial x}(n_i u_i) = 0. \tag{10.15}$$

In the momentum balance (Eq. (4.23)) for the ions in a non-magnetized plasma, the last term on the right-hand side can be neglected in case of a sufficiently large electric field. Then with the electrostatic potential $\Phi(x,t)$

$$\frac{\partial^2 x}{\partial t^2} = \frac{\partial u_i}{\partial t} = -\frac{e}{m_i}\frac{\partial}{\partial x}\Phi(x,t). \tag{10.16}$$

For the electrons, Boltzmann kinetics (Eq. (6.4)) is again assumed resulting in

$$n_e(x,t) = n_0 \exp\left(\frac{e\Phi(x,t)}{kT_e}\right), \tag{10.17}$$

where n_0 denotes the plasma density at the sheath edge. Finally, the Poisson equation must be fulfilled self-consistently

$$\frac{\partial^2 \Phi}{\partial x^2} = -\frac{e}{\varepsilon_0}\frac{\partial}{\partial x}(n_i - n_e). \tag{10.18}$$

With $x = 0$ at the electrode surface (see Figs. 6.2 and 6.3), a computational grid is extended to a sufficient depth x_p in the plasma. The initial conditions are $n_e(x,0) = n_i(x,0) = n_0$, where the presheath is neglected and thereby the factor $e^{-1/2}$ is omitted. With the voltage $-V_0$ applied at $t = 0$, the above system of equations is solved numerically with the boundary conditions $\Phi(0,t) = -V_0$ and $\Phi(x_p,t) = 0$.

Are result of such a calculation is given in Fig. 10.4, which shows the velocity of the sheath edge, which moves from the electrode into the plasma volume, vs. the position of the sheath edge. The results relies on experimental information about the plasma density and the electron temperature. It is in qualitative agreement with the analytical predictions of Section 6.4, with a high initial sheath edge velocity which tends to zero when approaching the static Langmuir-Child sheath. The prediction is in good agreement with experimental results obtained from time-resolved laser-induced fluorescence at varying distance from the electrode.

10.4 PARTICLE-IN-CELL COMPUTER SIMULATION

The most powerful plasma simulations rely on the so-called "particle-in-cell" (PIC) model. The plasma volume is subdivided by a two-dimensional or three-dimensional grid. All particles in the plasma are represented by "superparticles" of each species, which represent a large number of real particles of that species. The superparticles are allowed to move in the entire volume, whereas the fields which govern their motion are defined on the nodes of the grid. As indicated in Fig. 10.5, complicated configurations with superimposed electrical and magnetic fields can be treated in this way.

The procedure of a particle-in-cell simulation is shown schematically in Fig. 10.6. A computational loop is performed for sufficiently small time step Δt. Starting with the lower-left edge, the forces on each particles are calculated by interpolating the fields at the grid nodes j adjacent to the actual position x_i of the respective particle. Subsequently, the equation of motion can be solved for each particle. Particles arriving at the walls are treated according to the boundary conditions (loss or reemission at the walls) with a possible generation of secondary particles (secondary electron emission, wall sputtering, surface chemistry products, etc.). According to the individual collision cross sections and the corresponding collision probabilities, a Monte Carlo algorithm decides according if any particle undergoes a collision with any other species during the time interval. In case of a collision, the velocities and flight directions of the collision partners are revised. Then, the particle densities of different species can be determined at each grid node from the particle numbers around that grid node and the respective positions. Finally, the electrostatic potential and the electric field are obtained from the charged parti-

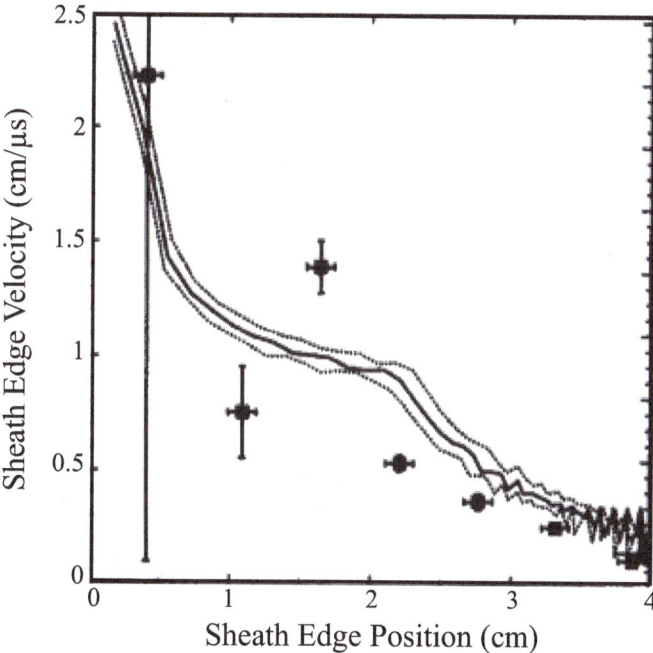

Figure 10.4: Fluid model simulation of the sheath edge dynamics after applying high-voltage DC bias, for a nitrogen plasma at a pressure of $6.7 \cdot 10^{-2}$ Pa with a plasma density of $6.25 \cdot 10^9$ cm^{-3} and an electron temperature of 0.57 eV. The applied bias voltage is -5 kV. The result (solid line) is compared to experimental data obtained by laser-induced fluorescence. The upper and lower dotted lines correspond to a variation of the plasma density by $\pm$ 10%. (From M.J. Goeckner et al. [12].)

cle densities at the grid nodes, using standard finite difference schemes for the solution of the Poisson equation, and the definition of the forces is reentered for the subsequent time step.

Results for a DC magnetron discharge (for the geometry, see Fig. 10.5) are shown in Fig. 10.7. Considerable inhomogeneities are obtained. Between the locations of maximum magnetic field, the electrostatic potential increases from the cathode potential in axial direction along a distance of about 1 cm, which is much larger than the width of the sheath. The Ar ion density shows a pronounced peak at the positions of maximum magnetic field and a diffusional tail. The sharp decrease closely behind the cathode results from a strong influence of collisions at the relatively high pressure. Neutral Cu atoms are generated by sputtering at the cathode. They are partly ionized in the region of high plasma density and form an ion distribution which is similar to the Ar ion distribution, at, however, a significantly lower concentration.

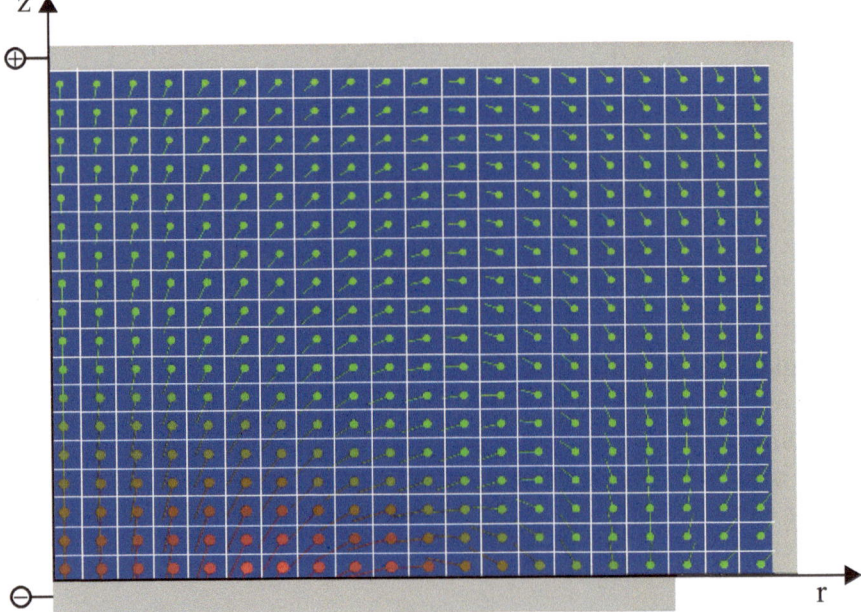

Figure 10.5: Particle-in-cell grid (white lines) of an axisymmetric plasme volume with a superimposed magnetic field (directional lines, color and diameter of points). r and z denote the cylindrical coordinates., the grey areas indicate the electrodes of a cylindrical magnetron configuration (see Section 11.2).

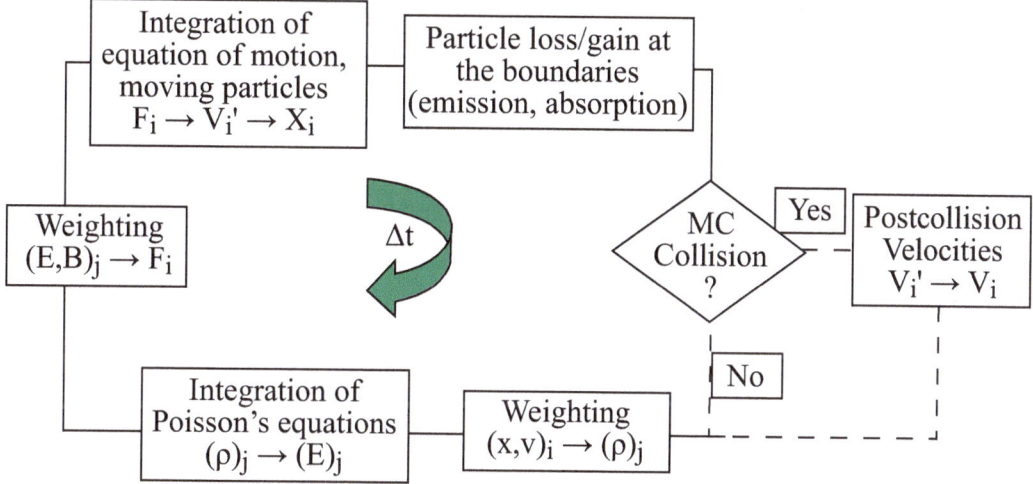

Figure 10.6: Computational scheme of a particle-in-cell plasma simulation. The index $i = 1 \ldots N_p$ denotes the individual superparticles, where the plasma is simulated by in total N_p superparticles of different species. $j = 1 \ldots N_g$ denotes the nodes of the discrete grid, with a total number N_g. x, υ, and ρ denote the superparticle position, velocity, and density, respectively, F the force acting on the superparticles, and E and B the electric and magnetic fields, respectively, at the grid nodes.

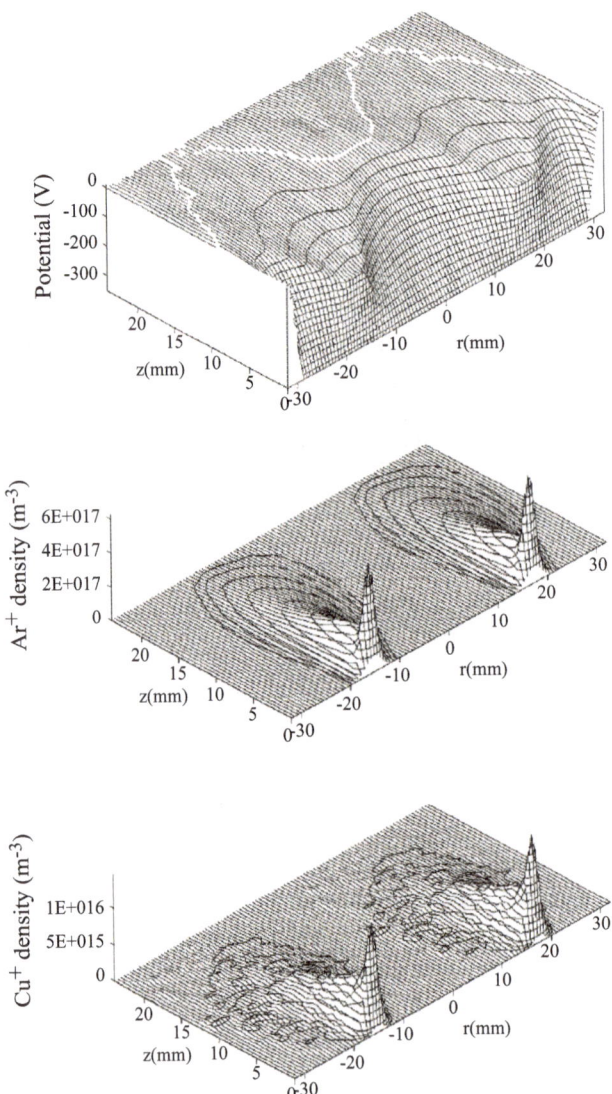

Figure 10.7: Electrostatic potential (top), Ar ion density (middle) and Cu ion density (bottom) in an Ar DC magnetron discharge (see Section 11.2), as obtained from a PIC computer simulation. The electrode configuration is shown in Fig. 10.5. $z = 0$ denotes the position of the cathode, the center position of the anode is at $z = 25$ mm. The maximum of the magnetic field is 0.12T at the position $(r, z) = (18$ mm, 0). The discharge pressure is 1.3 Pa, the discharge current 300 mA. In the top figure, the white line denotes the anode potential, $V = 0$. (From I. Kolev and A. Bogaerts [13].)

CHAPTER 11

Low-temperature DC Plasma

11.1 BREAKDOWN

The basic mechanism of plasma formation in an electric field is the ionization of neutral gas atoms, after electrons have gained sufficient energy between collisions to overcome the ionization threshold. This requires an initial generation of electrons, which could be artificially by, e.g., a field emission structure or by a heated filament, but also occurs naturally by, e.g., cosmic radiation.

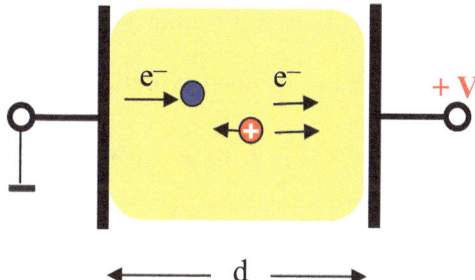

Figure 11.1: Electron multiplication in a DC discharge by ionizing collisions.

The formation of the discharge then depends on the multiplication of the initially generated electrons. This is described by the so-called first Townsend coefficient α, which denotes the relative increase in electron flux per unit path length,

$$dj_e = \alpha j_e \, dx \qquad (11.1)$$

α can also be interpreted as the number of ionization events electron and unit path length. Under the action of the electric field, the electrons are accelerated between the collisions. As shown above, this results in a Maxwellian which is "shifted" by the directed kinetic energy E_e. Accordingly,

$$\alpha = \frac{1}{\lambda_c} \exp\left(-\frac{U_i}{E_e}\right) \qquad (11.2)$$

with λ_c denoting the mean free path length for gas-kinetic collisions. The energy gain of the electrons along the mean free path length is given by the electric field E

$$E_e = \lambda_c e E. \qquad (11.3)$$

Then, with $\lambda_c \sim p^{-1}$ where p is the gas pressure, with constants A and B

$$\alpha = Ap \exp\left(-B\frac{pd}{V}\right),\tag{11.4}$$

where the electric field has been replaced by the applied voltage V and the distance d between the electrodes.

Without a continuous generation of primary electrons, the electron avalanche will stop after reaching the anode of the discharge. Such a discharge is called un-sustained. Examples are corona discharges for paper charging in copy machines or for exhaust cleaning in power plants or for chemical reactors.

Primary electrons which are necessary to sustain a "self-supported" discharge, can be continuously provided by the ion-induced production of secondary electrons at the cathode. The solution of Eq. (11.1) for a homogeneous discharge is

$$j_e(d) = j_e(0) \exp(\alpha d)\tag{11.5}$$

with $x = 0$ denoting the cathode position. As one ion is also produced in the cascade with each electron,

$$j_i(0) - j_i(d) = j_e(d) - j_e(0) = j_e(0)(\exp(\alpha d) - 1).\tag{11.6}$$

According to the definition of the SEEC (see Section 7.7)

$$j_e(0) = \gamma_e j_i(0).\tag{11.7}$$

When the charge multiplication is efficient, the ion flux at the anode can be neglected on the left-hand side of Eq. (11.6). From this, the breakdown condition results as

$$\alpha d = \log\left(1 + \frac{1}{\gamma_e}\right)\tag{11.8}$$

and, with Eq. (11.4)

$$V_b = \frac{Bpd}{\log(Apd) - \log\left(\log\left(1 + \frac{1}{\gamma_e}\right)\right)}\tag{11.9}$$

for the breakdown voltage. It depends on the product pressure and electrode distance. At lower pressure, the distance has to be larger. It varies weakly with the SEEC which depends on the cathode material.

The minimum voltage below which the discharge cannot be ignited at any pressure occurs at a pressure-distance product

$$pd|_{V_{\min}} = \frac{1}{A} \log\left(1 + \frac{1}{\gamma_e}\right).\tag{11.10}$$

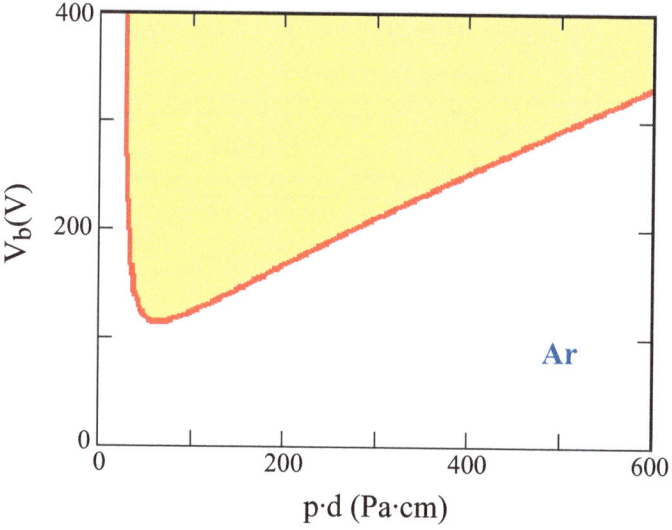

Figure 11.2: Paschen law of the breakdown voltage vs. the product of the pressure and the electrode distance, for $A = 1.02$ (Pacm)$^{-1}$ and $B = 17.7$ V/(Pacm), which are valid for Ar, and an SEEC of 0.1 which is typical for a DC discharge.

11.2 REGIMES OF OPERATION

As shown in Fig. 11.3, the optical appearance in DC discharge is rather inhomogeneous. According to Section 6.4, the main potential drop occurs at the cathode, where the ions are accelerated and the secondary electrons are produced. Above a dark space directly at the cathode, there is light emission due to collisional excitation by fast ions. The accelerated electrons cause the excitation in the so-called "negative glow," which is separated by two dark spaces. The more or less axially homogeneously emitting plasma toward the anode is called the "positive column," with a slightly increased light emission directly above the anode, where the ions are little accelerated by the floating potential.

This characterizes the regime of the glow discharge (see Fig. 11.4). If the glow discharge is pushed to higher currents, it enters the regime of abnormal glow. Here, the discharge becomes concentrated on a small cathode spot with a very high current density. Increasing the voltage in this regime gradually expands the cathode spot across the whole cathode area. The high plasma density and resulting low resistivity then results in an unstable regime where the voltage drops and the current increases drastically and the discharge enters the arc regime. Here, the thermionic emission of electrons from the cathode dominates the initial production of electrons. The emission current is given by the Richardson equation,

$$j_{e,th} = C T_c^2 \exp\left(-\frac{e \Phi_w}{k T_c}\right),$$ (11.11)

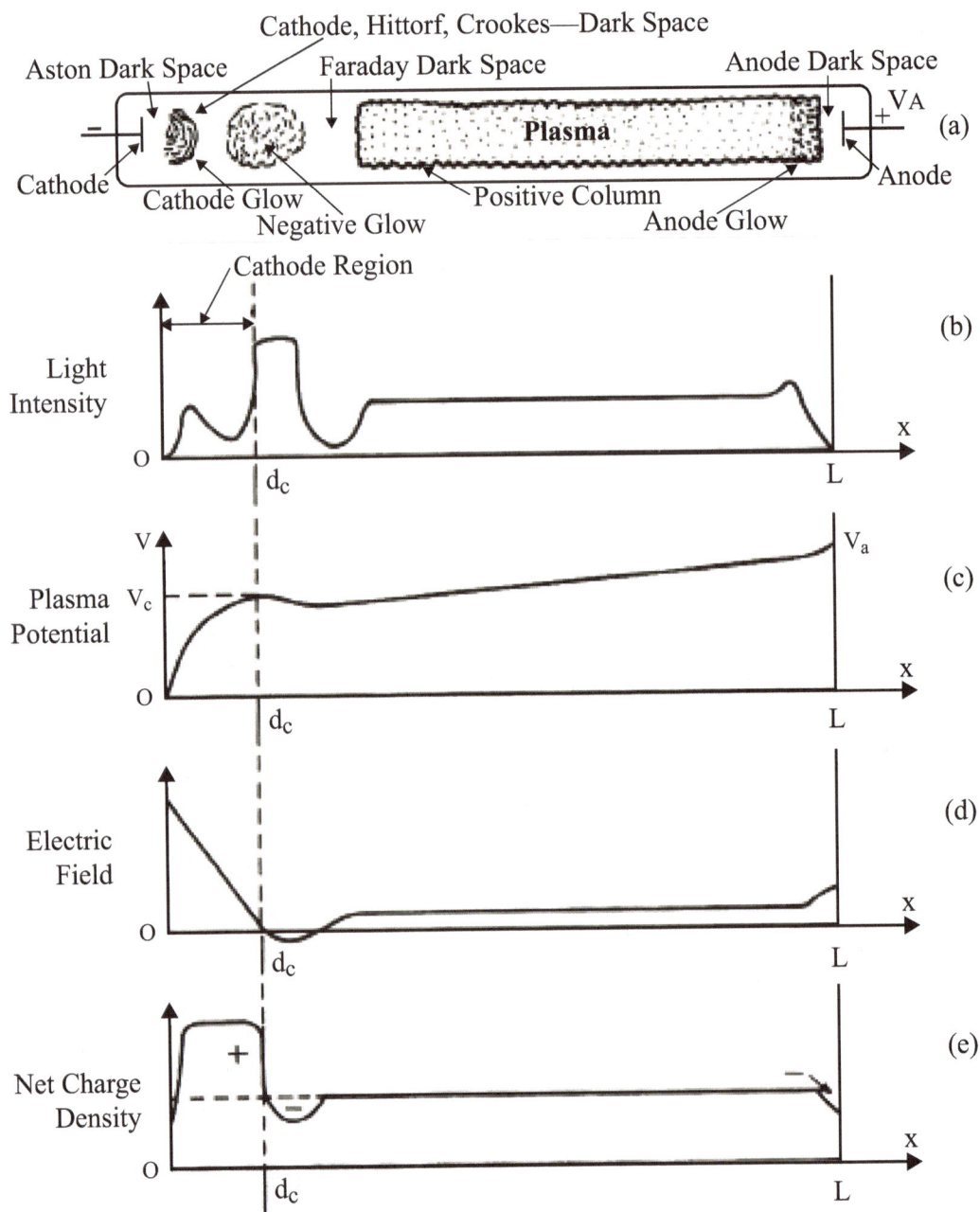

Figure 11.3: Appearance of a glow discharge and spatial variation of the light intensity, electro-static potential, electric field, and charge density.

where Φ_c denotes the work function and T_c the temperature of the cathode. Typical arc burning voltages are several 10 V. At sufficiently high current, the arc plasma becomes thermal.

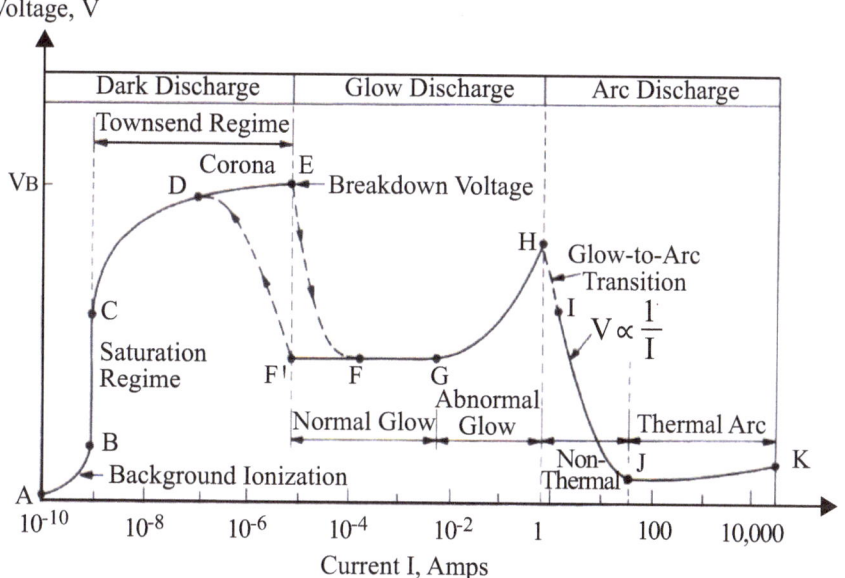

Figure 11.4: Regimes of the DC discharge. The breakdown voltage is given by the Paschen law, Eq. (11.9).

As stated above, the applied voltage mainly drops in the cathode fall, where the ions are accelerated to produce secondary electrons. Not only for this purpose, but also for the possibility to apply the DC discharge to surface treatment, it is interesting to look at the energy distribution of the ions arriving at the cathode. As indicated by Fig. 11.2, the DC discharge typically operates at pressures larger than 1 Pa where collisions become important. Here, the considerations of Section 6.5 apply.

11.3   DC MAGNETRON DISCHARGE

As mentioned above, DC discharges require rather high pressures. For certain applications, such as the widely used deposition of thin films by sputtering where neutral atoms are generated due to ion bombardment at the cathode and condensed on a substrate which is in the anode position, higher pressures are disadvantageous in three respects: (i) the ion energy at the cathode and thereby the sputtering yield is reduced (see Sections 7.4 and 11.1); (ii) the sputtered atoms are scattered away from the substrate so that the deposition rate is reduced; and (iii) the mean energy of the sputtered atoms (see Section 7.4) is reduced which deteriorates the film quality in many cases.

Therefore, a more efficient way of DC plasma generation at low pressures is required. This is realized by the magnetron configuration (see Fig. 11.5). Out of the many possible configurations. a planar cylindrical magnetron is shown in Fig. 11.5. A permanent magnet below the cathode, with a central inner pole and a ring shaped outer pole, generates a magnetic field in a geometry similar to a half-torus. Applying a DC voltage of a few 100 V results in a bright plasma ring above the cathode, which is embedded into a plasma of lower density. As the plasma remains essentially at floating potential, the potential drop across the cathode sheath corresponds to almost the full applied voltage as for the DC discharge. In the standard operation regime of a magnetron, the sheath is collisionless at a gas pressure of 0.1 Pa or below. Therefore, all ions hit the cathode with the full energy corresponding to the operation voltage. This increases not only the sputter yield (see Section 7.4), but also the production of secondary electrons. The electrons are accelerated back into the plasma and loose their energy by collisions with the gas atoms. The magnetic field confines the electrons in the plasma ring due to the pendulum motion in the inhomogeneous field, after the kinetic energy parallel to the field lines is reduced by collision.

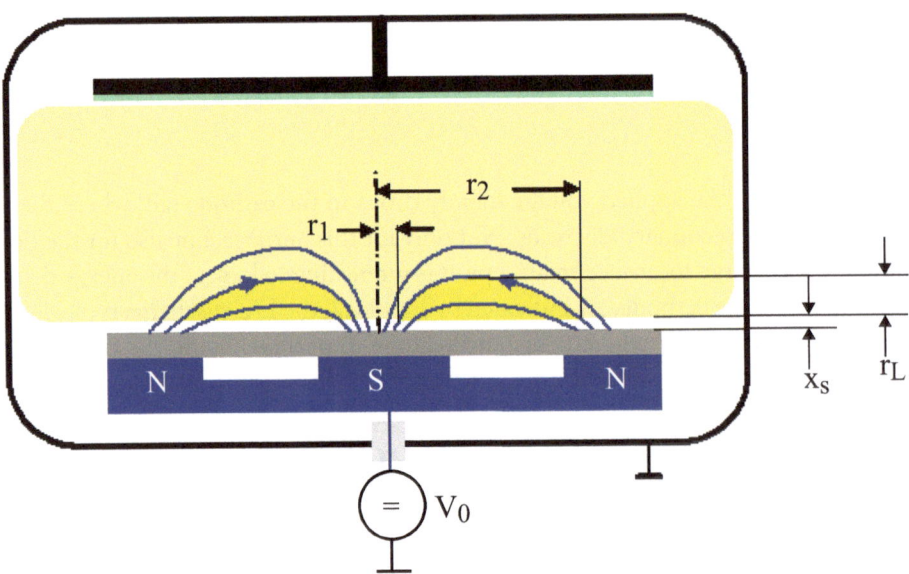

Figure 11.5: Scheme of the planar cylindrical DC magnetron discharge. The intensive plasma above the cathode ("target") surface is embedded in a low-density plasma filling the vessel. Atoms sputtered from the target by ion bombardment from the plasma (not shown) are deposited on the anode ("substrate") surface. The indicated geometrical quantities are discussed in the text.

As the electrons are accelerated in the sheath according to the applied voltage, each of them creates a number

$$N = \frac{eV_0}{U_{i,eff}} \approx \frac{eV_0}{2U_i} \tag{11.12}$$

of ion-electron pairs, when being slowed down in the plasma completely. Ui,eff denotes the average amount of energy spent per ionization event, as the electrons also loose energy by other processes than ionization. For simple estimations, it can be taken as two times the ionization potential of the gas atoms. However, some of the electrons are absorbed again at the cathode before loosing their full energy. This is accounted for by an effective secondary electron emission coefficient, which can be estimated by

$$\gamma_{e,eff} \approx \frac{\gamma_e}{2}. \tag{11.13}$$

Due to the applied field, all ions leave the plasma toward the target (cathode). In the stationary state, the number of ions produced in the plasma per unit of time is equal to the ion current I_i toward the target, i.e.,

$$I_i \gamma_{e,eff} N = I_i \tag{11.14}$$

or

$$\gamma_{e,eff} N = 1. \tag{11.15}$$

With Eqs. (11.12) and (11.13),

$$V_0 \approx \frac{4U_i}{e\gamma_e}. \tag{11.16}$$

For an SEEC of about 0.1 which is often assumed, and an ionization potential around 15 eV, this results in a burning voltage of about 600 V. However, the SEEC might be higher (see Section 7.7), and additional ionization by thermal electrons has been neglected above. Therefore, the burning voltage is lower that given by Eq. (11.16) by a factor of about 2, as found in experiments.

The gyroradius, which determines the axial extension of the plasma ring, is given by

$$r_{Le} = \frac{v_e}{\omega_{Ce}} = \frac{1}{B}\sqrt{\frac{2m_e V_0}{e}}. \tag{11.17}$$

For a typical magnetic field of 20 mT and a voltage of about 300 V this results in a Larmor radius of about 1 cm.

For an estimation of the width w of the plasma ring, the limiting field line is approximated by a circle segment with opening angle 2θ and radius R_c (see Fig. 11.6). From geometry,

$$\sin \vartheta = \frac{w/2}{R_c} \tag{11.18}$$

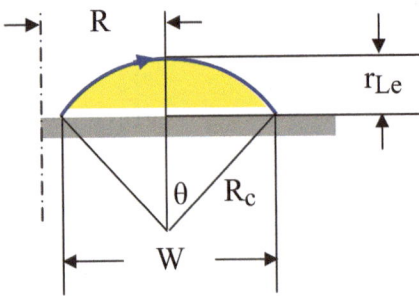

Figure 11.6: Simplified geometry of the magnetron discharge.

and

$$r_{Le} + R_c \cos \vartheta = R_c. \tag{11.19}$$

Combining Eqs. (11.18) and (11.19) yields

$$w = 2\sqrt{r_{Le}(2R_c - r_{Le})}. \tag{11.20}$$

With a radius of curvature of typically 2 cm for a smaller laboratory magnetron the resulting width is about 3 cm.

As the ions are essentially not influenced by the magnetic field (their Larmor radius is in the order of 1 m), the average ion flux to the substrate is given by the Child-Langmuir law, Eq. (6.36). On the other hand, it is given by

$$j_i = \frac{I_i}{2\pi e R w} \tag{11.21}$$

as the current is only determined by the ion flux to the cathode. With a typical current of about 1 A, the resulting average current density is about 10 mA/cm². Using the Bohm flux of Eq. (6.12) and assuming a typical electron temperature of 3 eV for a low-pressure discharge, a plasma density of $n_e \approx 5 \cdot 10^{11}$ cm^{-3} is obtained, which represents a very high density for a low-pressure plasma.

CHAPTER 12

Low-temperature RF Plasmas

12.1 CAPACITIVELY COUPLED RF DISCHARGE

If an AC voltage is applied to a plasma in the configuration of Fig. 12.1, the plasma behaves just as an alternating DC plasma provided the frequency of the applied voltage is smaller than the ion plasma frequency (see Eq. (6.30)),

$$\omega < \omega_{pi}. \tag{12.1}$$

For low-pressure plasmas, the ion plasma frequency is in the order of 10 MHz. Thereby, the frequency regime of RF plasmas is above several MHz. A typical frequency is the industrial band of 13.56 MHz, but also higher frequencies up to about 100 MHz are applied. The arrangement for the capacitively coupled discharge (Fig. 12.1) is very similar to that for the DC discharge. RF is applied to the "hot" electrode through a network matching to the combined impedance of the electrode configuration and the plasma, which includes a blocking condensor to decouple any DC potential.

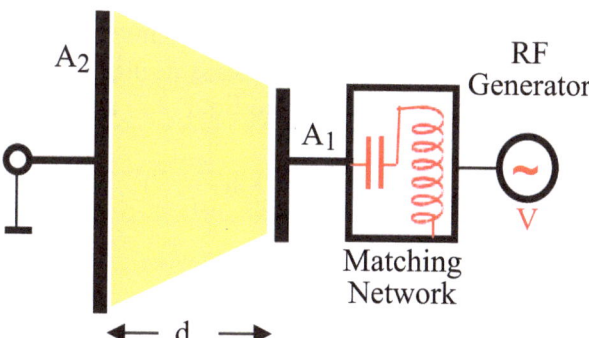

Figure 12.1: Scheme of a capacitively coupled RF discharge with the areas A1 of the powered electrode and A2 of the grounded electrode, and the interelectrode distance d.

Typical RF amplitudes are around 500 V. Thus, during the RF oscillations, the electrodes form in an alternating and antiphase way a biased sheath. As the RF frequency is in the order of the ion plasma frequency or higher, the ions do not entirely follow, and the sheath widths oscillate to some extent around an average thickness. In Fig. 12.1, the plasma will move to the right during the positive half-wave of the RF voltage being applied to the hot electrode 1, and to the left during the negative half-wave.

Assuming in the simplest approximation an ion matrix sheath (see Section 6.4) with a constant ion density n and zero electron density, the electric field in the sheath result from the Poisson equation according to

$$E(x,t) = \frac{en}{\varepsilon_0}(x - s(t)) \tag{12.2}$$

with the geometry of Fig. 12.2 and the assumption that the field at the plasma boundary is negligible.

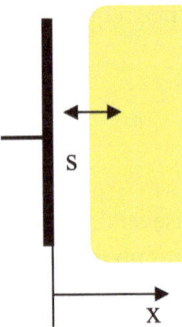

Figure 12.2: Oscillating plasma boundary under the influence of an RF field.

The RF current $I(t)$ flows through the serial arrangement of the two sheaths and the plasma. In the sheaths, it is largely dominated by the displacement current, while the conduction current by the ion and electron flow to the electrodes is negligible. The displacement current density is given by the Maxwell equations and from Eq. (12.2) according to

$$j(t) = \varepsilon_0 \frac{dE}{dt} = -ne\frac{d}{dt}s(t) \tag{12.3}$$

and the current with the electrode area A

$$I(t) = -neA\frac{d}{dt}s(t) = I_{RF}(t) = -I_0 \cos \omega t, \tag{12.4}$$

where I_0 denotes the RF current amplitude. The phase has been chosen for the reason of later convenience. By integration,

$$s(t) = \bar{s} + s_0 \sin \omega t \tag{12.5}$$

with the integration constant $\bar{s}$, which represents the time-averaged position of the plasma boundary, and the amplitude

$$s_0 = \frac{I_0}{ne A\omega}. \tag{12.6}$$

Although the conduction current is small, it has to cancel in average as no net DC current is flowing due to the blocking condensor. In the picture of the ion matrix sheath, this can only be achieved by a momentary complete collapse of the sheath so that electrons can escape from the plasma, so that

$$\bar{s} = s_0. \tag{12.7}$$

The voltage drop across the sheath can then be calculated by integrating Eq. (12.2)

$$V(t) = -\int_0^{s(t)} E(x,t)\,dx = \frac{en}{2\varepsilon_0}(s(t))^2 \tag{12.8}$$

and with Eqs. (12.5) and (12.7)

$$V(t) = \frac{en}{2\varepsilon_0}s_0^2(1 + 2\sin\omega t + \sin^2\omega t). \tag{12.9}$$

In the scheme of Fig. 12.1, the time-dependent voltage between the plasma and the (grounded) electrode 2 becomes

$$V_{p2}(t) = \frac{I_0^2}{2\varepsilon_0 ne\,A_2^2\omega^2}(1 + 2\sin\omega t + \sin^2\omega t). \tag{12.10}$$

On the (powered) electrode 1, the phase is shifted by 180° (a maximum sheath width at one of the electrodes occurs with a minimum at the other one). Therefore,

$$V_{p1}(t) = \frac{I_0^2}{2\varepsilon_0 ne\,A_1^2\omega^2}(1 - 2\sin\omega t + \sin^2\omega t). \tag{12.11}$$

As these voltages are measured in opposite directions, they add to the total applied RF voltage $V_{RF}(t)$, according to

$$V_{RF}(t) = V_{p2}(t) - V_{p1}(t). \tag{12.12}$$

For a symmetric arrangement with $A_1 = A_2 = A$, the harmonic RF voltage results consistently as

$$V_{RF}(t) = V_0 \sin\omega t = \frac{2I_0^2}{\varepsilon_0 ne\,A^2\omega^2}\sin\omega t, \tag{12.13}$$

where V_0 denotes the RF amplitude. The results of Eqs. (12.10), (12.11), and (12.13) are displayed in Fig. 12.3.

The sheaths act as rectifiers. A mean potential of the plasma, $< V_p >$, is established with respect to both electrodes which is somewhat less than half the RF amplitude.

To describe the Ohmic heating by RF, the power density can be written as

$$P_{RF} = \frac{1}{2}\eta_{RF}\,j_0^2, \tag{12.14}$$

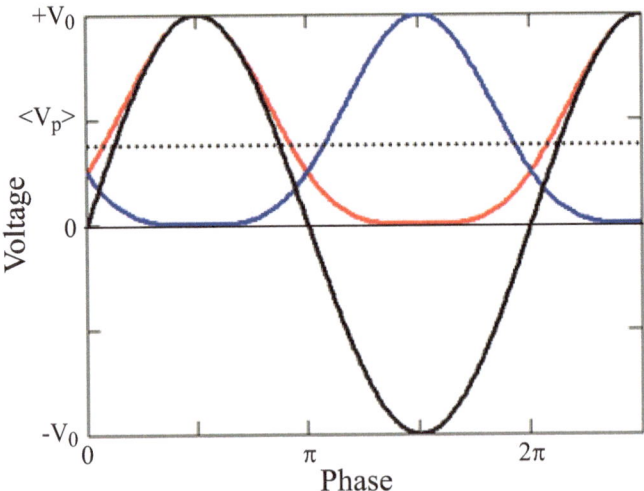

Figure 12.3: Voltages for a capacitively coupled RF plasma (see Fig. 12.1) in symmetric configuration: Voltage between the plasma and electrode 2 (red line), voltage between the plasma and electrode 1 (blue line), and RF voltage applied to electrode 1 (black line). The dotted line denotes the average voltage between the plasma and both electrodes, which is 0.375 V_0.

where η_{RF} is the resistivity according to Eq. (5.56) and j_0 the amplitude of the RF current density. Integrating over the plasma length yields with $j_0 = I_0/A$ yields the power dissipated per unit area

$$P_{RF} = \frac{1}{2}\eta_{RF}I_0^2\frac{d}{A^2}. \tag{12.15}$$

From Eqs. (12.13) and (5.55),

$$P_{RF} = \frac{m_e}{4}d\frac{\varepsilon_0}{e}\nu_{ce}\left(1 + \frac{\omega^2}{\nu_{ce}^2}\right)V_0\omega^2 \tag{12.16}$$

which, surprisingly, scales linearly with the applied voltage and inverts the frequency dependence of Eq. (5.55).

Assuming for a typical low-pressure plasma a pressure of 1 Pa and an electron temperature of ~ 2 eV, a corresponding electron collision cross section of $\sim 5 \cdot 10^{-15}$ cm^2 (see Fig. 12.4), the mean electron collision frequency (Eq. (2.5)) is about 100 MHz. Thus, at the standard RF frequency of 13.56 MHz corresponding to $\omega = 85$ MHz, the term in brackets is not significantly different from 1 and varies slowly with ω.

A further heating mechanism in RF plasmas is the so-called "stochastic" heating, which is due electron acceleration in the field of the moving boundary. As shown schematically in Fig. 12.4, electron from the plasma are reflected by the boundary as usually, but feel here the momentum by the moving boundary. A net gain of energy is only possible if the phase of the electron motion with respect to the RF field is changed, just as by collisions with atoms as for the Ohmic heating.

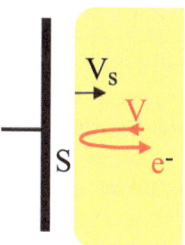

Figure 12.4: Stochastic heating of an electron with the initial velocity v by the moving plasma boundary with velocity v_s (schematic).

An electron with initial velocity v from the plasma receives after the collision with the "moving wall" with velocity v_s the velocity

$$v' = -v + 2v_s. \tag{12.17}$$

If $f_s(v, t)$ denotes the velocity distribution of electrons at the sheath edge as a function of time t, the number of electrons which collide with the moving boundary per unit area and unit time is

$$d\dot{N}_{es} = (v - v_s) f_s(v, t)\, dv. \tag{12.18}$$

which results in a differential power transfer per unit area

$$dP_{st} = \frac{m_e}{2}(v'^2 - v^2)(v - v_s) f_s(v, t) dv. \tag{12.19}$$

Inserting Eq. (12.17) and integrating over velocity, the power transfer per unit area becomes

$$P_{st} = -2m_e \int_{v_s}^{\infty} v_s (v - v_s)^2 f_s(v, t) dv. \tag{12.20}$$

In the simplest approximation and in consistency with the matrix sheath description, the time dependence of f_s is neglected. Further, we assume for most of the plasma electrons $v \gg v_s$ so that the lower integration limit can be set to zero. Setting $v_s = v_{s0} \cos \omega t$ and averaging over time yields

$$P_{st} = 2m_e v_{s0}^2 \int_0^{\infty} v f_s(v) dv. \tag{12.21}$$

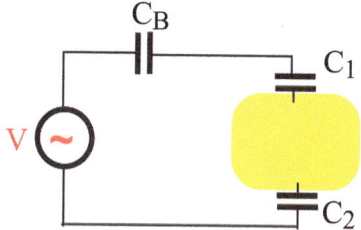

Figure 12.5: Equivalent circuit for a capacitively coupled RF plasma with the blocking condensor C_B and the sheath capacitors C_1 and C_2.

Assuming an equal plasma density at both plasma boundaries, the scaling between the sheath thickness and the voltage across the sheath is according to the Child-Langmuir law, Eq. (6.36)

$$x_{si} \sim \langle V_{pi} \rangle^{3/4} \tag{12.29}$$

which results in an exponent of $n = 4$ in Eq. (12.27). However, it is questionable to what extent the Child-Langmuir sheath is established during the short RF half period (see Section 6.4). Therefore, the discrepancy can be considered to be due to the simplifications involved in both of the above models. Experiments show an exponent of $n = 2, \ldots 3$ which is between the above predictions, and is also consistent with a collisional sheath which would predict an exponent of $n = 2.5$ (see Section 6.5).

Combining Eqs. (12.12) and (12.27), the time-averaged voltage at the powered electrode becomes

$$\langle V_1 \rangle = \langle V_{p2} \rangle - \langle V_{p1} \rangle = -\langle V_{p1} \rangle \left(1 - \left(\frac{A_1}{A_2} \right)^n \right). \tag{12.30}$$

In the standard situation with $A_1 < A_2$, this results in a negative DC "self-bias" voltage at the powered electrode. In the limit $A_1 \ll A_2$, $< V_{p2} >$ becomes very small so that the DC voltage at the powered electrode, which is of the order of the RF amplitude V_0, becomes practically identical to the DC potential between the powered electrode and the plasma.

12.2 ION ENERGY DISTRIBUTION

The time-varying potential between the plasma and the electrodes provides ion bombardment of the electrode surfaces. For a simplified description of the ion energy distribution, we remain in the picture of the ion matrix sheath. According to Eqs. (12.5), (12.6), (12.10), and (12.13), the time-dependent sheath thickness and voltage across the sheath can be written as

$$s(t) = s_0 (1 + \sin \omega t) \tag{12.31}$$

and

$$V(t) = \frac{ne}{2\varepsilon_0} s_0^2 (1 + \sin \omega t)^2 = \frac{V_0}{4} (1 + \sin \omega t)^2 \tag{12.32}$$

with the time averages $< s > = s_0$ and $< V > = 3/8 \cdot V_0$. Solving the equation of motion for a singly charged ion starting at the average plasma boundary position in an approximate average electric field $E = < V > /s_0$, results in an average transit time to the electrode of

$$\langle t_i \rangle = \sqrt{\frac{8}{3} \frac{1}{\omega_{pi}}}. \qquad (12.33)$$

In a low-temperature RF plasma, typical electron densities are around 10^{10} cm^{-3} corresponding to ion plasma frequencies around 10 MHz. Also, the RF frequency is in this range so that the ion transit time is in the order of the RF period. Therefore, the ion energy distribution at the electrode results both from the oscillation of the plasma boundary and the dynamics of the ion in the time-dependent RF field. In view of the severe simplification involved in the ion matrix picture, it would be inadequate to solving this rather complicated problem. Instead, we will treat two extreme cases. The distribution function of ion energies E_i at the electrode is given by

$$f_e(E_i) = C \frac{dN_i}{dE_i} = C \frac{dN_i}{dt} \frac{dt}{dV(t)} \frac{dV(t)}{dE_i}, \qquad (12.34)$$

where C is a normalization constant, N_i denotes the number of ions impinging on the electrode, and t the time at which an ion starts from the plasma boundary. For a constant plasma density the number of ions per unit of time is constant.

First, we assume that the ion transit time is infinitely short, i.e., $\omega_{pi} >> \omega$. Then, the ion energy distribution just reflects the distribution of the voltage across the sheath, which is given by the middle differential in Eq. (12.34) (the last term is equal to e^{-1}). Calculating t explicitly from Eq. (12.32) and differentiating with respect to $V = E_i/e$ yields

$$f_e(E_i) \propto \frac{1}{\sqrt{1 - \left(\sqrt{\frac{4E_i}{eV_0}} - 1\right)^2}} \frac{1}{\sqrt{E_i e V_0}}; \qquad \omega_{pi} >> \omega. \qquad (12.35)$$

For $\omega_{pi} << \omega$, each ion is accelerated by the time-varying electric field in the sheath. For a simple solution, we approximate the field according to (see Eqs. (12.31) and (12.32))

$$E(t') \approx \frac{V(t')}{s(t')} = \frac{n e s_0}{2\varepsilon_0}(1 + \sin \omega t'). \qquad (12.36)$$

Solving the corresponding equation of motion with the initial conditions $x = s(t)$ and $dx/dt = 0$ for $t' = 0$, calculating the transit time t_i' from $x = 0$, and evaluating for $t_i' >> \omega^{-1}$ results in a velocity at the electrode

$$\dot{x}(t_i') = -\omega_{pi} s_0 \sqrt{1 + \sin \omega t} \qquad (12.37)$$

from which the kinetic energy of the ion becomes as function of its start time t

$$E_i(t) = \frac{eV(t)}{1 + \sin \omega t} = \frac{e}{2}\sqrt{V(t)V_0}. \qquad (12.38)$$

From this, the last term in Eq. (12.34) is evaluated resulting in

$$f_E(E_i) \propto \frac{1}{\sqrt{1 - \left(\sqrt{\frac{4E_i}{eV_0}} - 1\right)^2}} \frac{1}{V_0}; \qquad \omega_{pi} << \omega. \qquad (12.39)$$

The ion energy distributions (IED's) resulting from Eqs. (12.35) and (12.39) are shown in Fig. 12.6. Basically, the distributions peak at the full applied RF voltage and zero, reflecting the probability of residence of the voltage across the sheath. The asymmetry of the voltage function (see Fig. 12.3) results in an asymmetry of the distribution toward low energy for transient times being short compared to the RF period. In the inverse case, ions emitted from the boundary at large instantaneous sheath thickness (and thereby voltage across the sheath) are more accelerated than those emitted from the shallow sheath boundaries, which causes a shift toward higher energies. Thus, higher RF frequencies result in higher average ion energies.

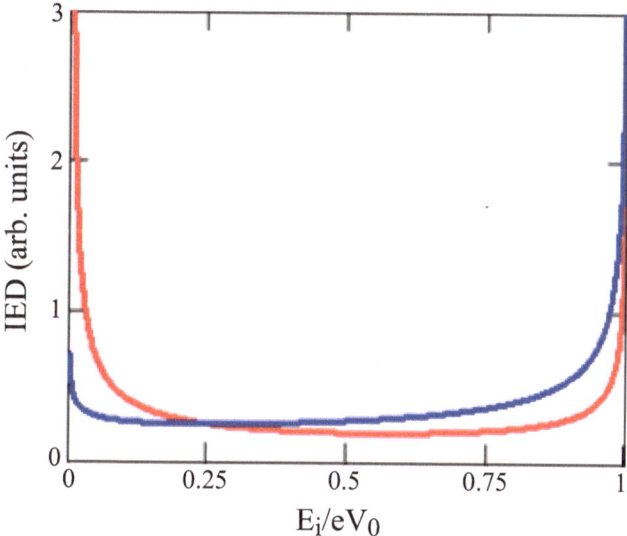

Figure 12.6: Ion energy distributions from an RF plasma in simple matrix sheath approximation, for RF frequencies being small (red) and large (blue) compared to the ion plasma frequency.

More realistic treatments of the RF plasma sheath without the matrix sheath approximation result in considerably smaller oscillations of the plasma boundary around the mean position. Then, for $\omega_{pi} >> \omega$, the IED reflects the oscillation of the boundary as above, but with a much lower amplitude. In this regime, it is independent of the ion mass. For $\omega_{pi} \approx \omega$, the ions perform a pendulum motion during the acceleration toward the electrode, the amplitude of which depends on the ion mass. For light ions, the ion energy can even exceed the RF amplitude. An example of a corresponding measurement is given in Fig. 12.7.

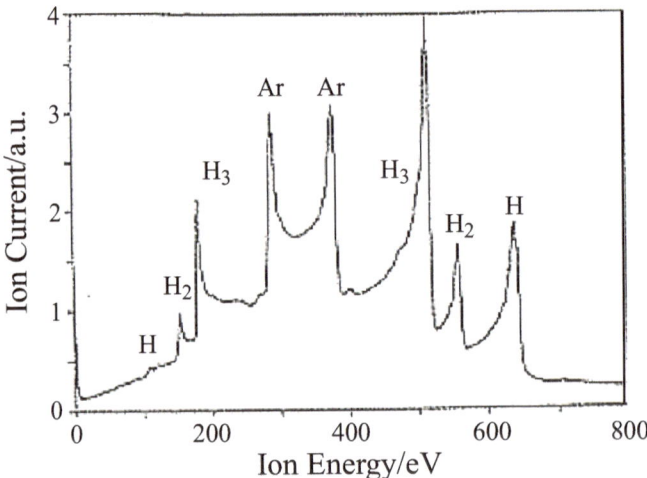

Figure 12.7: Experimental ion energy distribution from an $Ar+H_2$ RF plasma, in the regime $\omega_{pi} \approx \omega$. Heavier ions are more inert with respect to the oscillating electric field. Experimental parameters: RF frequency 13.56 MHz, RF amplitude 465 V, pressure $4 \cdot 10^{-3}$ mbar, $Ar/H_2 = 0.4$. (From D. Field et al. [14].)

For $\omega_{pi} << \omega$, all ions are unable to follow the RF field, so that the IED becomes rather narrow around the mean sheath voltage, and again independent of the ion mass.

The IED becomes even more complicated when the sheath is collisional. In addition to the RF oscillations, the mean collision frequency has a significant influence. This may lead to highly structured IEDs even for monatomic gases as shown in Fig. 12.8. Here, the measured IED corresponds very well to a theoretical model.

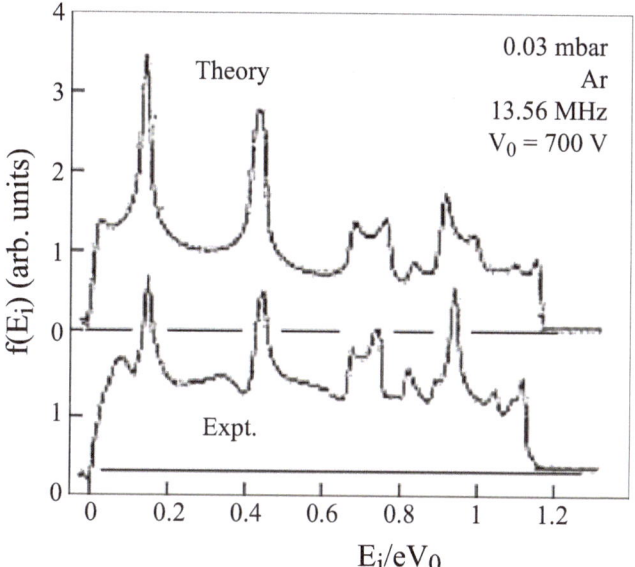

Figure 12.8: Collision-dominated ion energy distribution from an Ar RF plasma, from experiment and model calculation. Due to the relatively high pressure and the relatively large RF amplitude corresponding to a large average sheath thickness, the sheath becomes collisional. (From C. Wild and P. Koidl [15].)

Integrating for a one-dimensional Maxwellian velocity distribution results in

$$P_{st} = \frac{m_e}{2} v_{s0}^2 n v_e,$$
(12.22)

where v_e denotes the thermal velocity of the electrons in the plasma. In the plasma, the RF current is essentially by conduction, which yields relative to the moving boundary

$$I_0 = -n e v_{s0} A.$$
(12.23)

Using Eq. (12.13), the stochastic heating power per unit area results as

$$P_{st} = \frac{m_e}{4} v_e \frac{\varepsilon_0}{e} V_0 \omega^2$$
(12.24)

which, at sufficiently small frequency, exhibits the same dependencies on the RF voltage and frequency as the Ohmic heating, Eq. (12.16), but is independent on the collision frequency. In comparison with Ohmic heating and the typical conditions discussed there,

$$\frac{P_{st}}{P_{RF}} \approx \frac{v_e}{d v_{ce}}.$$
(12.25)

For an electrode separation of ~ 10 cm this ratio becomes ~ 0.1, indicating that stochastic heating is less important at pressure of 1 Pa and above, but might become dominant at pressures below 0.1 Pa.

Applying Eq. (12.12) to the non-symmetric case,

$$V_{RF}(t) = \frac{I_0^2}{\varepsilon_0 n e \omega^2} \left(\frac{1}{A_1^2} + \frac{1}{A_2^2} \right) \sin \omega t + \frac{I_0^2}{2\varepsilon_0 n e \omega^2} \left(\frac{1}{A_2^2} - \frac{1}{A_1^2} \right) (1 + \sin^2 \omega t)$$
(12.26)

which is no longer purely harmonic and demonstrates the limitation of the simple matrix sheath description. However, the second term is small compared to the first one provided the difference of the areas is not too large. Neglecting this problem, Eqs. (12.10) and (12.11) demonstrate that the amplitudes of the voltages, or the mean voltages across the sheaths, scale with the electrode areas according to

$$\frac{\langle V_{p1} \rangle}{\langle V_{p2} \rangle} = \left(\frac{A_2}{A_1} \right)^n$$
(12.27)

with the exponent $n = 2$. Alternatively, the plasma and the sheaths under the influence of the (12.27), RF voltage can be modeled by a simplified circuit according to Fig. 12.5. The capacity of the sheath condensors is given by the electrode areas and by the thickness of the sheaths. Then for $i = 1, 2$

$$\frac{\langle V_{pi} \rangle}{\langle I \rangle} = \frac{1}{\omega C_i} \sim \frac{x_{si}}{A_i}.$$
(12.28)

CHAPTER 13

Magnetic Confinement Nuclear Fusion Plasma

Early development of plasma physics was mainly promoted by the aim to apply controlled nuclear fusion to energy generation. For this purpose, the ion temperature of the plasma has to be sufficiently large to overcome the Coulomb barrier and to obtain a significant rate of nuclear fusion reactions. Therefore, the ion confinement time has to be sufficiently large. A self-sustained nuclear fusion plasma, which remains stationary only due to energy delivered by the fusion reaction, has not been realized on earth so far. Recent research, however, allows to extrapolate present machine date toward an ignited plasma, which shall be demonstrated in the international ITER project.

13.1 FUSION REACTIONS

The presently most promising reaction for controlled nuclear fusion is the deuterium-tritium reaction between the heavy isotopes of hydrogen

$$^2H + ^3H = D + T \Rightarrow ^4He(3.52 MeV) + n(14.07 MeV) \tag{13.1}$$

assuming thermal reaction partners. The Q value of 17.59 MeV of the reaction is distributed into the given kinetic energies of the reaction partners according to the reaction kinematics.

The basic principle of energy production is to employ the fast neutrons, which penetrate the walls of the plasma container, for the generation of heat by moderation (like in a fission reactor), whereas the resulting α particles would deliver their energy mainly to the plasma in order to sustain the discharge.

Deuterium is available in practically unlimited amounts and easily to recover, e.g., from the sea water. Tritium will be generated in the fusion device using the breeding reaction

$$^6Li + n \Rightarrow ^3H + ^4He \tag{13.2}$$

by thermalized neutrons in the so-called blanket material around the plasma vessel. Again, ^6Li is available in minerals in sufficiently large quantities to guarantee a long-term world-wide energy supply. In terms of radiation hazard, the tritium handling has to be strictly controlled, but the reaction product of both above reactions, helium, is entirely unproblematic, which represents the biggest advantage against nuclear fission.

The rate coefficient of the fusion reaction, Eq. (13.1), is shown in Fig. 13.1 vs. the plasma temperature.

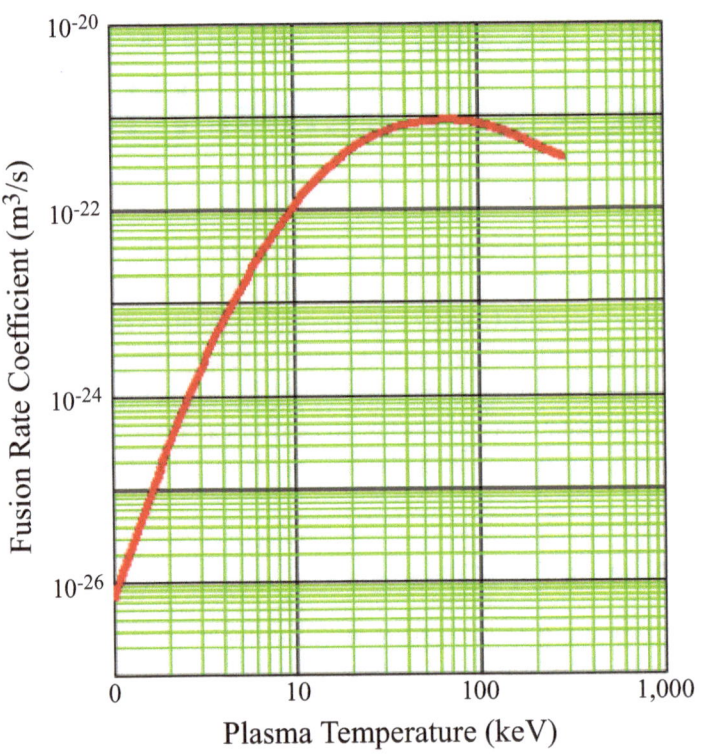

Figure 13.1: Rate coefficient of the DT nuclear fusion reaction in a thermal plasma as function of its plasma temperature.

13.2 IGNITION

For a fusion plasma, which is self-sustained only by the power delivered by the fusion reaction, the simplified global balance of energy density reads

$$\frac{n_e \cdot 3k T_e}{\tau_E} = \langle \sigma v \rangle_f n_D n_T \varepsilon_f, \tag{13.3}$$

where τ_E denotes the energy confinement time, $<\sigma v>_f$ and ε_f the fusion rate coefficient and energy gain per reaction, respectively, and n_D and n_T the deuterium and tritium densities, respectively. As the plasma is assumed to be thermal and fully ionized, each electron adds its own kinetic energy and the one of the corresponding ion on the left-hand side. Ideally, the deuterium and tritium densities are equal:

$$n_D = n_T = \frac{n_e}{2}.$$ (13.4)

Therefore, the simplified condition for the ignition of the self-sustained fusion plasma becomes

$$n_e \tau_E = \frac{12 k T_e}{\langle \sigma \upsilon (k T_e) \rangle_f \varepsilon_f}$$ (13.5)

which neglects energy losses from the plasma by other than charged particle transport. This defines a minimum product of electron density and energy confinement time at given electron temperature. Evaluating Eq. (13.5) for a temperature between 10 and 20 keV, the corresponding rate coefficient according to Fig. 13.1, and the initial energy of the α particles (see Eq. (13.1)) yields the Lawson criterion

$$n_e \tau_E > 10^{14} \, \text{s/cm}^3 \quad \text{at} \quad k T_e > 10 \, \text{keV}.$$ (13.6)

The most prominent additional channel of energy loss form the plasma is the escape of bremsstrahlung, which arises from the Coulomb deflection of the electron trajectories during collisions with the ions. The power density of bremsstrahlung is given by

$$p_{bs} = \frac{16 \alpha^3 \hbar^2}{\sqrt{3} m_e^{3/2}} n_e^2 \sqrt{k T_e} Z_{eff}.$$ (13.7)

Here, α denotes the fine structure constant. The effective (ion) charge Z_{eff} is defined by

$$Z_{eff} = \frac{1}{n_e} \sum Z^2 n_Z,$$ (13.8)

where the sum extends over all ion species with charge Z and density n_Z. The effective charge takes into account possible impurities in the plasma, which mainly arise from erosion of the vessel walls by energetic ion bombardment. It is seen that in particular heavy impurities increase the bremsstrahlung loss significantly. For example, a 1% impurity of an ion with an (assumed) charge of 50 will increase the power loss by a factor of 25.

The refinement of the balance Eq. (13.3) then reads

$$\frac{n_e \cdot 3 k T_e}{\tau_E} = n_e^2 \left(\frac{\langle \sigma \upsilon \rangle_f \varepsilon_f}{4} - \frac{16 \alpha^3 \hbar^2}{\sqrt{3} m_e^{3/2}} \sqrt{k T_e} Z_{eff} \right)$$ (13.9)

from which a modified ignition criterion (see Eq.(13.5)) can be derived. Fig. 13.2 displays the results of the simple criterion and including bremsstrahlung. It is seen in a pure hydrogen plasma bremsstrahlung does not alter the conditions significantly in the desired regime around the curve minimum, whereas $Z_{eff} = 10$ puts already severe restrictions, in particular requiring an increased temperature, which is difficult to achieve.

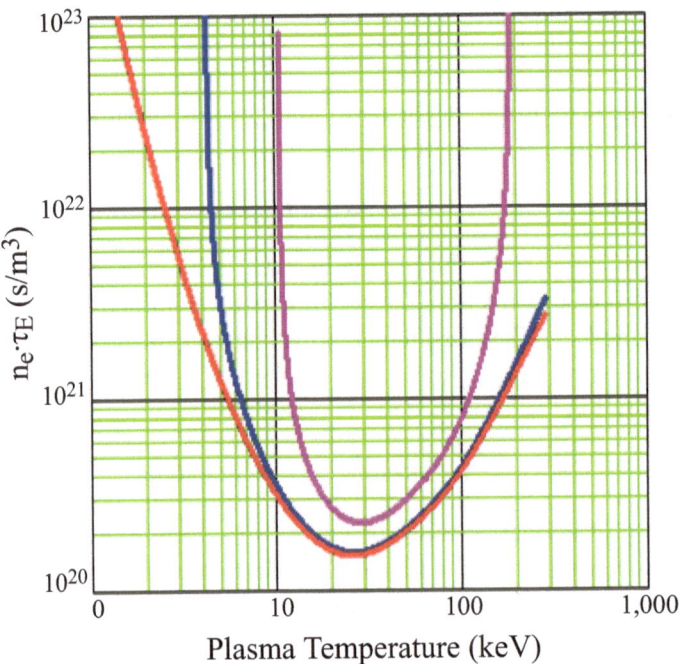

Figure 13.2: Ignition criterion according to the simple energy balance (Eq. (13.5)) (red line) and including bremsstrahlung with $Z_{eff} = 1$ and 10 (blue and magenta lines, respectively). Ignition takes place in the parameter range above the curves.

13.3 MACHINE CONCEPTS

The extremely high temperatures which are necessary for nuclear fusion cannot be achieved without efficient mechanisms of plasma confinement. Without such measures, the electron temperature would directly result from the geometrical dimensions of the vessel (see Eq. (10.6)), with a siginificant increase in size resulting only in a small change of the temperature due to the strongly varying ionization rate coefficient. An efficient means to provide additional confinement is to embed the plasma into a magnetic field which is parallel to the walls of the vessels, as charged particle would be attached to the field lines and only get lost by the strongly retarded collisional cross-field diffusion (see Section 5.4).

The ideal plasma container would be a sphere as it minimizes the surface-to-volume ratio and thereby the losses to the wall. However, there can be no closed magnetic field configuration on the surface of a sphere which is a direct consequence of the Maxwell equations.

A simple arrangement is the axisymmetric magnetic field which is generated by a series of parallel coils. In order to achieve the required $n_e \tau_E$ product (see Section 13.2), either the density or the confinement times can be increased. The former possibility was realized in so-called pinch discharges which make use of the adiabatic compression of a cylindrical plasma at strongly varying magnetic fields (see Section 3.7). The extremely high densities which can be achieved in this way result in a high β value (see Section 4.3). Such plasmas tend to become unstable.

Alternatively, magnetic mirror configurations have been exploited (see, e.g., Fig. 3.7). In addition to instabilities, these devices suffer from the end losses as the "bottleneck" field still allows the fastest electrons and ions to escape.

Consequently, a "closed cylindrical" configuration is promising, which is realized by a toroidal geometry. As shown in Fig. 13.3, a corresponding toroidal magnetic field can be realized by a set of coils which are arranged vertically around the centerline of the device.

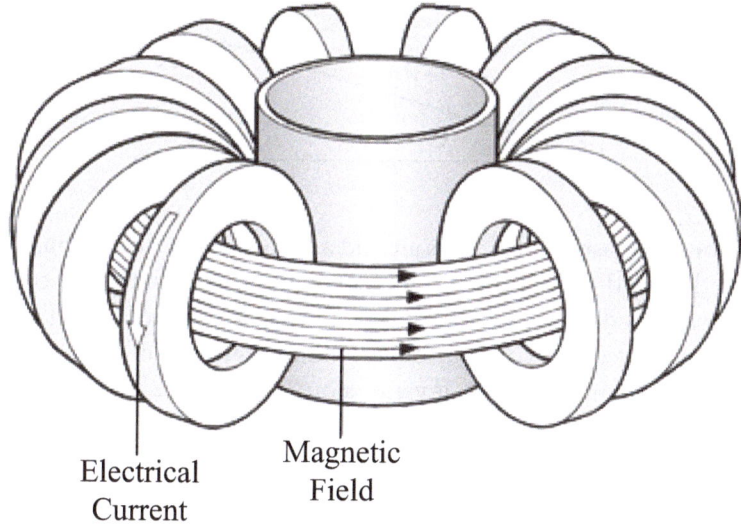

Electrical
Current

Magnetic
Field

Figure 13.3: Toroidal plasma device with the magnetic field generated by a set of electromagnetic coils.

Figure 13.4 shows a section with the geometry and the magnetic field indicated. As a disadvantage, the magnetic field is not constant across the cross section of the plasma. Due to the arrangement of the coils, it decreases $\sim R^{-1}$ in outward direction. Consequently, the $B \times \nabla B$ drift (see Section 3.4) separates charges of different polarity and creates and electric field in vertical direction. This causes in turn an outward ExB drift so that the plasma will be pushed against the outer wall. This is depicted in Fig. 13.5.

The ExB drift would be avoided by providing a poloidal component of the magnetic field by means of a sheared field (see Fig. 13.6), which distributes the charged particle over all poloidal positions so that the charge separation is counteracted.

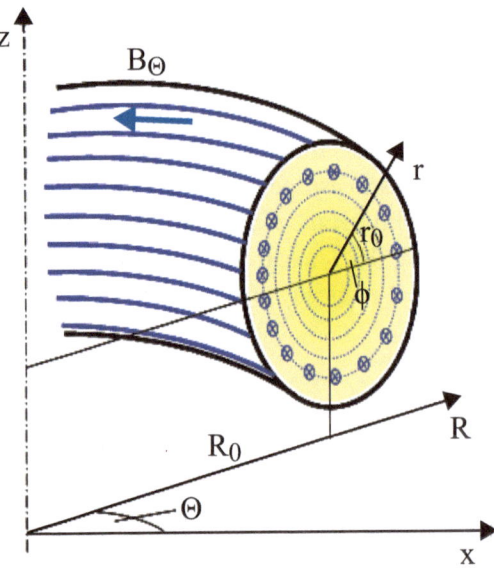

Figure 13.4: Section of a plasma-filled torus around a center line in z direction, with the toroidal geometry indicated. R and θ denote the toroidal coordinates, r and ϕ the coordinates in the poloidal plane, R_0 and r_0 denote the large and small radius, respectively. The toroidal field B_θ is shown at the outer surface. The concentric rings indicate surfaces of constant pressure. As the magnetic field decreases at increasing R, it varies along the surfaces of constant pressure.

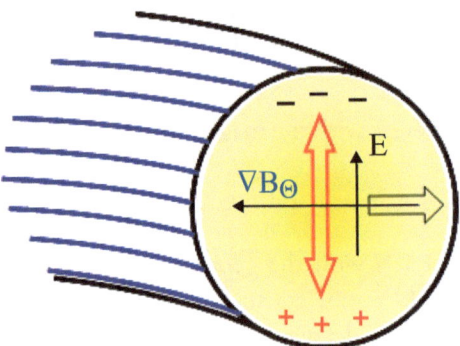

Figure 13.5: Fields an drifts in a toroidal plasma. The red and black bold arrows indicate the $B \times \nabla B$ and ExB drifts, respectively.

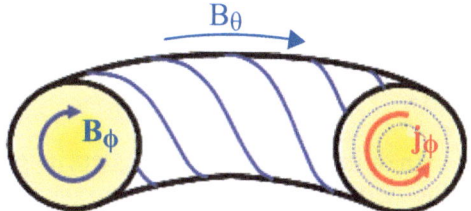

Figure 13.6: Sheared magnetic field by superimposing a poloidal field component B_ϕ. The diamagnetic current j_ϕ is also indicated.

For the sheared field, any field line is closed, if sufficiently many circulations N_c both toroidally and poloidally are considered. Correspondingly, a **safety factor** Q is defined as the ratio of the toroidal to the poloidal circulations of a closed field line, i.e.,

$$Q = \lim_{N_c \to \infty} \frac{N_\theta}{N_\phi}, \qquad (13.10)$$

where N_θ and N_ϕ denote the number of toroidal and poloidal circulations, respectively, of the closed field line. The nomenclature is due to the fact that the plasma becomes unstable both for too small and too large Q. The acceptable range of Q results in a field topology with small field inclination.

In the fluid picture, the toroidal plasma consists of toroidal shells of constant density (also indicated in Figs. 13.5 and 13.6) with increasing density toward the centerline. Neglecting collisions, the charged particle are attached to the field lines and perform a spiral motion in toroidal direction and along the poloidal circles. As the particles are transported in both directions depending on their initial velocity, there is no net current. However, the radial density gradient requires a diamagnetic current, j_ϕ, according to Eq. (4.36). If the magnetic field shear is small, this poloidal current is perpendicular to the field lines. Then, according to Eq. (4.33), stable operation requires

$$\vec{j}_\phi \times \vec{B} = \vec{\nabla} p = \vec{C} onst \qquad (13.11)$$

as the density gradient is constant in the shell. However, as B varies radially and as there are no current sources within the plasma, Eq. (13.11) is violated, which leads to fluctuations and instability. This can be resolved by adding a net current in toroidal direction, which is essentially parallel to the field lines at small field inclination. This current can be written as

$$\vec{j}_\theta = \frac{\vec{j} \cdot \vec{B}}{B^2} \vec{B}, \qquad (13.12)$$

where j denotes the total current. The requirement for a source-free plasma volume then reads

$$\vec{\nabla} \cdot (\vec{j}_\phi + \vec{j}_\theta) = 0 \qquad (13.13)$$

or, using Eq. (4.36),

$$\frac{1}{B^2}(\vec{B} \times \vec{\nabla} p + (\vec{j} \cdot \vec{B})\vec{B}) = \vec{C}onst \tag{13.14}$$

which requires a deliberate design of the local current (i.e., depend on the radial position of each shell) which has to be consistent with the magnetic field and the density gradient.

Conveniently, the toroidal current can be employed to generate the required shear of the magnetic field. There are different ways to generate the toroidal current. Today, the most advanced concept is the tokamak, which is sketched in Fig. 13.7. The plasma acts as the secondary winding of an axial transformer. Poloidal field coils are arranged around the vertical center line of the device, which are supplied with a current which varies linearly with time so that a constant poloidal magnetic field is generated. As the current variation cannot be sustained for infinite time, the tokamak operates inherently in pulsed mode. This is a disadvantage as, e.g., it imposes severe conditions of thermal cycling to the surrounding materials, even if the nowadays envisaged pulse times are longer than 5 min.

A concept for continuous operation is the stellerator, where the sheared magnetic can be produced additional spiral coils around the plasma vessel. Due to extreme forces, however, this would not be feasible for fusion device. A modern configuration employs distorted toroidal coils which provide poloidal field components in connection with a periodically changing plasma shape. This is shown in Fig. 13.8. The state of development of stellarators, however, is presently far behind that of tokamaks, so that ignition cannot be expected for the forthcoming generation of devices.

13.4 TRANSPORT

The following section is related to the tokamak. The **classical transport** of charged particles from the plasma center toward the wall given by cross-field ambipolar diffusion as described in Section 5.4. This mechanism is dominant for sufficiently large collision frequencies.

At low collision frequency, the transport of the particles is mainly governed by the magnetic field topology. Following the lines of the sheared field, the poloidal motion meets subsequently high-field and low-field regions. As for the magnetic bottle configuration described in Section 3.8, this may lead to a pendulum motion for sufficiently low velocities $v_=$ parallel to the field. A projection of this motion onto a poloidal plane is shown in Fig. 13.8.

A half period of the motion corresponds to a fraction $\phi_b/2\pi$ of a full poloidal circle where ϕ_b is the opening angle in the poloidal plane. This transform into the fractional passage of a toroidal circle via the safety factor, Eq. (13.10). As the field inclination is small, multiplying with $2\pi R_0$ yields the length of trajectory of the gyrocenter per half period,

$$L_b \approx QR_0\phi_b. \tag{13.15}$$

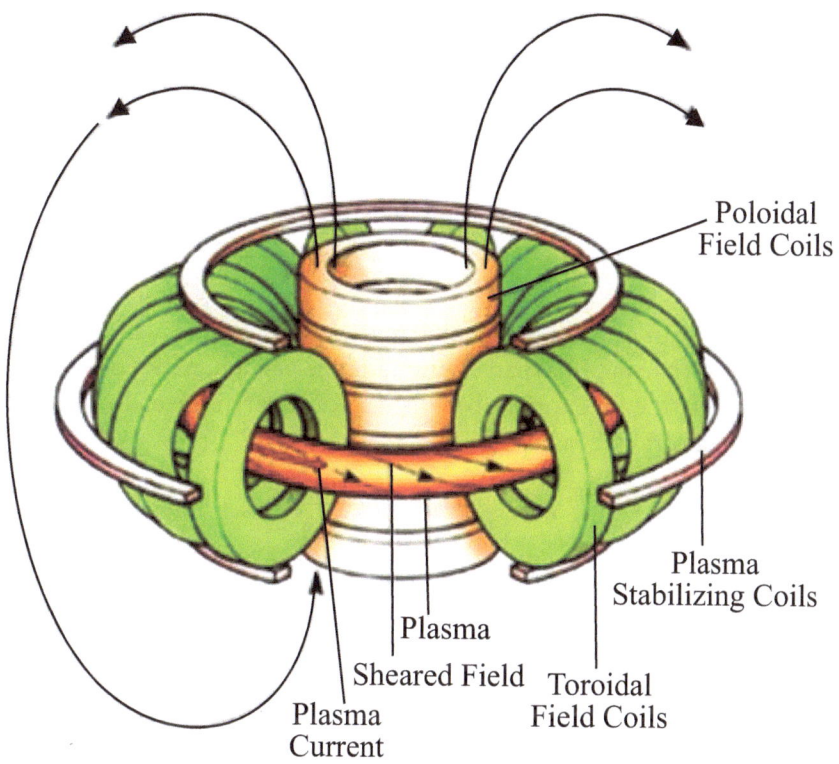

Figure 13.7: Tokamak device with some poloidal field lines indicated.

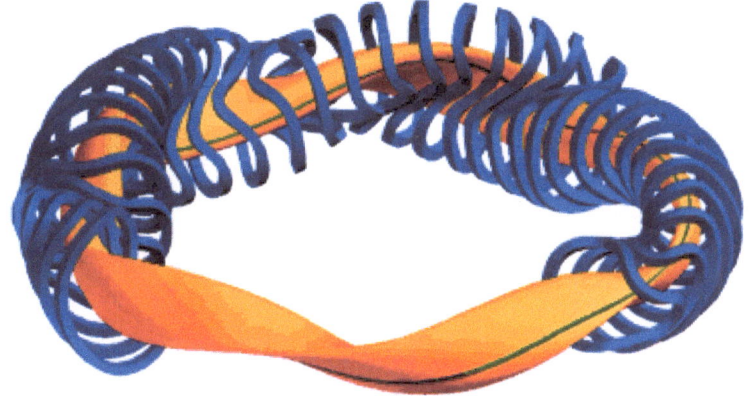

Figure 13.8: Magnetic coil arrangement and plasma shape of the W7-X project at Greifswald, Germany.

Both the $B \times \nabla B$ and the curvature drift act on these particles in identical vertical direction. As $B \sim 1/R$, the gradient results as

$$\nabla B = -\frac{B}{R} \approx -\frac{B}{R_0}. \tag{13.16}$$

From Eqs. (3.15) and (3.19), the combined drift velocity v_D results then as

$$v_{Db} = \frac{m}{eBR_0}\left(v_=^2 + \frac{1}{2}v_\perp^2\right) \approx \frac{m}{2eBR_0}v_\perp^2. \tag{13.17}$$

It can be seen that this drift causes a broadening in the poloidal plane, which, together with the toroidal motion, yields a banana-like spatial contour of the pendulum motion, which is therefore called the **banana orbit**. The time for one banana half-trajectory is given by $t_b = L_b/v_=$. Thus, the width of the banana orbit becomes

$$w_b = v_{Db}t_b = r_L Q\frac{v_\perp}{v_=}\phi_b \tag{13.18}$$

in case of collisions, the width of the banana orbit then becomes the characteristic cross-field jump distance instead of the Larmor radius. A representative width is given for $\phi_b \approx \pi$. Therefore, the coefficient of **neoclassical diffusion** can be estimated to

$$D_{nc} \approx w_b^2 v_c = \left(\pi r_L Q\frac{v_\perp}{v_=}\right)^2. \tag{13.19}$$

Neoclassical diffusion is typically 100 times faster than classical cross-field diffusion.

References

[1] F. F. Chen, *Introduction to Plasma Physics*, Plenum Press, New York, 1974. DOI: 10.1007/978-1-4757-0459-4.

[2] R. J. Goldston and P. H. Rutherford, *Introduction to Plasma Physics*, Institute of Physics Publ., Bristol, 1995. DOI: 10.1201/9781439822074.

[3] C. K. Birdsall and A. B. Langdon, *Plasma Physics Via Computer Simulation*, Institute of Physics Publ., Bristol, 1991. DOI: 10.1887/0750301171.

[4] M. A. Lieberman and A. J. Lichtenberg, *Principles of Plasma Discharges and Materials Processing*, John Wiley & Sons Inc., New York, 1994. DOI: 10.1002/0471724254.

[5] R. Hippler, S. Pfau, M. Schmidt, and K. H. Shoenbach, Eds., *Low Temperature Plasma Physics*, Wiley-VCH, Berlin, 2001.

[6] B. N. Chapman, *Glow Discharge Processes—Sputtering and Plasma Etching*, John Wiley & Sons, New York, 1980. DOI: 10.1063/1.2914660.

[7] D. E. Post and R. Behrisch, Eds., *Physics of Plasma Wall Interactions in Controlled Fusion*, NATO ASI Series B, vol. 131, Plenum Press, New York, 1986. DOI: 10.1007/978-1-4757-0067-1.

[8] S. Glasstone and R. H. Loveberg, *Kontrollierte Thermonukleare Reaktionen*, Verlag Karl Thiemig KG, München 1964.

[9] K. Miramoto, *Plasma Physics and Controlled Nuclear Fusion*, Springer, Berlin, 2005. DOI: 10.1007/3-540-28097-9.

[10] R. E. H. Clark and D. H. Reiter, *Nuclear Fusion Research: Understanding Plasma-surface Interactions*, Springer, Berlin, 2005. DOI: 10.1007/b138970.

[11] L.E. Kline, W.D. Parlow, and W.E. Bies, Electron and chemical kinetics in methane RF glos-dischard deposition plasmas, *Journal of Applied Physics*, 65, 70, 1989. DOI: 10.1063/1.343378. 116

[12] M.J. Goeckner and S.M. Malik, Laser-induced fluorescence measurement of the dynamics of a purlse planar sheath, *Physics of Plasmas*, 1, 1064, 1994. DOI: 10.1063/1.870924. 119

[13] I. Kolev and A. Bogaerts, PIC – MCC Numerical simulation of a DC Planar Magnetron, *Plasma Processes Polymers*, 3, 127, 2006. DOI: 10.1002/ppap.200500118. 122

[14] D. Field, D.F. Klemperer, P.W. May, and Y.P. Song, Ion energy distributions in radio-frequency discharges, *Journal of Applied Physics*, 10, 82, 1991. DOI: 10.1063/1.350247. 140

[15] Ch. Wild and P. Koidl, Structured ion energy distribution in radio frequency glow-discharge systems, *Appl. Phys. Lett.*, 54, 505, 1989. DOI: 10.1063/1.100913. 141

Author's Biography

NIKOLAS XIROS

Nikolas Xiros is a J.L. Goldman Endowed Professor in Naval Architecture and Marine Engineering at the University of New Orleans. His career spans more than 20 years in both industry and academia and his expertise lies within the fields of marine and electromechanical systems engineering. He holds an electrical engineer's degree, a marine engineering doctorate, and a MS in mathematics. His research interests are process modeling and simulation, system dynamics, identification and control, reliability, signal and data analysis. He is author of many technical papers as well as other academic texts and reports.